Swacch India

Published by
ACADEMIC FOUNDATION
in association with

Centre for Economic and Social
Studies (CESS), Hyderabad

K.S. Babu is an Associate Professor at Centre for Economic and Social Studies (CESS), Hyderabad. He has MA in Anthropology from Andhra University, Visakhapatnam and PhD from University of Delhi, Delhi. He has 25 years of research experience in the fields of healthcare, health insurance, NGOs, rural development, tribal development etc. He has published articles in professional journals. He has co-edited (with S. Mahendra Dev) two books: *India: Some Aspects of Economic and Social Development* and *India's Development: Social and Economic Disparities*. His current research focuses on medical tourism, water and sanitation, and tribal issues.

M. Gopinath Reddy is Professor of Political Science and Public Policy at the CESS, Hyderabad and also Principal Coordinator for the Division for Sustainable Development Studies (DSDS) at the CESS. He was Principal Coordinator for the recently concluded Research Unit for Livelihoods and Natural Resources (RULNR), supported by Jamsetji Tata Trust (JSTT, Mumbai) and brought a number of monographs/ working papers under forest ecosystem spanning across major forest states of India. He has published various articles relating to forest governance and livelihoods and natural resources in the journals of *Development and Change, The Small Scale Forestry Journal, Economic and Political Weekly* and written chapters in the book titled: *Forests, People and Power: The Political Ecology of Reform in South Asia*. He has written two books: *Political Economy of Watershed Management: Policies, Institutions, Implementation & Livelihoods* (co-authored with Ratna Reddy and John Soussan) and *A Study of Redistributive Politics and its Impact on PRIs: A Case of Assembly Constituency Development Programme (MLA LADS Scheme in Andhra Pradesh).*

S. Galab, a Development Economist, is currently Director, CESS, Hyderabad, and is also leading the Young Lives—an International Longitudinal Project in Andhra Pradesh, on childhood poverty. He has carried out research in the broad field of rural development on various aspects like rehabilitation and resettlement of project affected people, natural resource management, solid waste management in urban areas, handloom and power loom sector, agrarian distress and farmers' suicides, rural indebtedness, MGNREGS, dry land agriculture. He has also worked on the effectiveness of public policies related to poverty alleviation including social capital, women empowerment, child labour and livelihoods. He was a member of Farmers Welfare Commission appointed by the Government of Andhra Pradesh in response to the agrarian crisis in the united state of Andhra Pradesh. He is presently a member of the Commission on "Sustainable Agriculture Development of Andhra Pradesh". He is one of the co-editors of the books: *On Joint Forest Management in Andhra Pradesh*; *Organising Poor Women: The Andhra Pradesh Experience* and *Centre-State Relations in Indian Fiscal Context*.

P. Usha is a Project Research Associate who worked for the project "Study on Water and Sanitation Programmes and Health Status of the Communities: A Study of Three Indian States, Madhya Pradesh, Odisha and Andhra Pradesh" at the CESS, Hyderabad. She did MA in Population Studies from Annamalai University and has worked on reproductive and child health.

Swacch India

Experience of Water and Sanitation Programmes in Three Indian States of Madhya Pradesh, Odisha and Andhra Pradesh

K.S. Babu, M. Gopinath Reddy,
S. Galab and P. Usha

ACADEMIC FOUNDATION
NEW DELHI

www.academicfoundation.org

First published in 2017
by

ACADEMIC FOUNDATION
4772-73 / 23 Bharat Ram Road, (23 Ansari Road),
Darya Ganj, New Delhi - 110 002 (India).
Phones : 23245001 / 02
Fax : +91-11-23245005.
E-mail : booksaf@gmail.com
www.academicfoundation.org

in association with
Centre for Economic and Social Studies (CESS), Hyderabad.

Disclaimer:
The findings/views/opinions expressed in this book are solely those of the authors and do not necessarily reflect the views of the publisher.

Cataloging in Publication Data--DK
Courtesy: D.K. Agencies (P) Ltd. <docinfo@dkagencies.com>

Babu, K. S. *(Professor of anthropology)*, **author.**
Swacch India : Experience of Water and Sanitation Programmes in Three Indian States of Madhya Pradesh, Odisha and Andhra Pradesh / K.S. Babu, M. Gopinath Reddy, S. Galab and P. Usha.
pages cm
Includes bibliographical references.
ISBN 9789332704206

1. Water-supply--India--Madhya Pradesh. 2. Water-supply--India--Odisha. 3. Water-supply--India--Andhra Pradesh. 4. Sanitation--India--Madhya Pradesh. 5. Sanitation--India--Odisha. 6. Sanitation--India--Andhra Pradesh. 7. Water--Health aspects--India--Madhya Pradesh. 8. Water--Health aspects--India--Odisha. 9. Water--Health aspects--India--Andhra Pradesh. I. Gopinath Reddy, M., author. II. Galab, Shaik, author. III. Usha, P. (Researcher in population studies), author. IV. Title.

TD303.M33B33 2017 DDC 363.6109543 23

Typeset, printed and bound by The Book Mint, New Delhi.
www.thebookmint.in

Contents

List of Tables and Figures 9

Foreword 23

Acknowledgements 25

1. **Introduction** 27

Rationale Underlying the Research Study

Water and Sanitation

Review of Literature

Programmes Related to Water and Sanitation

Major Policy Initiatives with respect to Water, Sanitation and Health

Relevance of the Study

2. **Methodology and Sampling Design** 53

Rationale Underlying the Sample Frame

Limitations

3. **Water and Sanitation Programmes and their Effects on the Health Status of Communities in Madhya Pradesh** 67

Introduction

Water and Sanitation

A Profile of the Study Villages in the District

A Profile of Households and Children's Behaviour Related to Water, Sanitation and Hygiene

VWSCs and VHSCs in the Study Villages

An Analysis of Information Based on Child Health and Hygiene

Summary of Focus Group Discussions (FGDs) Held in Both NGP and Non-NGP Villages in MP

Summary

4. Water and Sanitation Programmes and their Effects on the Health Status of Communities in Odisha. 125

Introduction

Water and Sanitation

A Profile of the Study Villages in the District

A Profile of Households and Children Behaviour Related to Water, Sanitation and Hygiene

An Analysis of Information Based on Child Health and Hygiene

Illness Episodes

Summary of Focus Group Discussions (FGDs) Conducted in both the NGP and Non-NGP Villages in Odisha

Summary

5. Water and Sanitation Programmes and their Effects on the Health Status of Communities in Andhra Pradesh 183

Introduction

Water and Sanitation

A Profile of the Study Villages in the District

A Profile of Households and Children Behaviour Related to Water, Sanitation and Hygiene

VWSC and VHSC in the Study Villages: Awareness Level Regarding VWSC and VHSC among the Local Population

A Profile of Diseases Among Adults

Summary of Discussions Based on Focus Group Discussions (FGDs) in Both NGP and Non-NGP Villages in AP

Summary

6. A Comparison of Findings across the States of Madhya Pradesh, Andhra Pradesh and Odisha 241

Main Sources of Water across the Three States
Quantity of Water Supply
Quality of Water
Water Related Hygiene Practices
Practices of Hygiene Maintenance
Treating of Drinking Water
Type of Housing Facility
Sanitation Facility
Frequency of Cleaning Toilets
Constraints Related to the Use of In-House Toilet Facility
Practices Related to the Disposal of Children's Excreta
Personal Hygiene Practices
Functioning of the Local Level Institutions
Morbidity among Children
Morbidity among Adults
Summary and the Way Forward

References 261

LIST OF TABLES AND FIGURES

TABLES

1.1 Incentive Pattern under NGP: A Unique Experiment in Sanitation Coverage in India 41

1.2 State-Wise Break Up of Award Winning (NGP) *Gram Panchayats* 41

2.1 Selection and Sample Design 54

2.2 Sampled Villages and Households across the Three States 55

2.3 NGP Villages and Non-NGP Villages in Visakhapatnam District in AP 60

2.4 NGP Villages and Non-NGP Villages in Dhar District in Madhya Pradesh 61

2.5 NGP Villages and Non-NGP Villages in Odisha 62

2.6 Study Matrix 62

2.7 Research Tools Developed and Used for the Study 65

3.1 Demographic Profile of Madhya Pradesh 68

3.2 Situation of Drinking Water and Sanitation in MP. 70

3.3 Gender and Literacy Profiles of the Respondents in NGP and Non-NGP Villages: MP. 78

3.4 Profile of Households by Type of House in NGP and Non-NGP Villages: MP. 79

3.5 Distribution of the Sample Households by Type of Toilet in NGP and Non-NGP Villages: MP 80

3.6 Respondent's Perceptions Related to the Non-Use of Toilets in NGP and Non-NGP Villages: MP 81
3.7 Respondents' Perceptions Related to the Use of Toilets, by Educational Level in NGP and Non-NGP Villages: MP 82
3.8 Details of Hand Wash Habit After Defecation in NGP and Non-NGP Villages: MP 83
3.9 Water Availability and Usage (by Source) in NGP and Non-NGP Villages: MP 84
3.10 Perceptions of the Respondents Regarding the Quality of Water in NGP and Non-NGP Villages: MP 85
3.11 Details of Adequacy/Inadequacy of Water Availability for Daily Needs: MP 85
3.12 Perceptions of the Respondents Regarding the Adequacy of Water Supply during the Past One Year in NGP and Non-NGP Villages: MP 86
3.13 Particulars of Season-Wise Water Insufficiency (in terms of Intensity): MP 86
3.14 Information on the Storage Capacity (Drinking Water) of Water Tanks/Drums/Vessels Across the Sample Households in NGP and Non-NGP Villages: MP 87
3.15 Information on Practices of Drawing Drinking Water from Containers in NGP and Non-NGP Villages: MP 88
3.16 Distribution of Households by Frequency of Cleaning Water Containers Used for Drinking Water in NGP and Non-NGP Villages: MP 89
3.17 Information on Treatment of Drinking Water Before Consumption, by Educational Level of the Respondents in NGP and Non-NGP Villages: MP. 90
3.18 Details of Measures Taken for Treating Water Before Consumption by Sample Households in NGP and Non-NGP Villages: MP 91

3.19. Information on the Major Problems Related to Water Supply in the Sample Villages (NGP and Non-NGP): MP 92

3.20 Distribution of Respondents Reporting Major Problems Regarding Water Supply in NGP and Non-NGP Villages: MP. 93

3.21 Status of Water Supply in ICDS Centres in NGP and Non-NGP Villages: MP. 93

3.22 Respondents' Awareness Regarding the Advantages of Using In-House Latrine/Toilet/Lavatory Facility in NGP and Non-NGP Villages: MP 94

3.23 Respondents Encounter Problems Regarding the Use of Toilets in NGP and Non-NGP Villages: MP 94

3.24 Problems Encountered by the Households Relating to Toilet Use in NGP and Non-NGP Villages: MP 95

3.25 Information on the Practices of Disposing of Children's Stools Post Defection in NGP and Non-NGP Villages: MP 96

3.26 Information on Hand Wash Facility and Type of Hand Wash Near Toilets: MP 97

3.27 Frequency of Cleaning Toilets in NGP and Non-NGP Villages: MP 97

3.28 Treatment Seeking Behaviour of the Households with Respect to Dehydration among Children in NGP and Non-NGP Villages: MP. 98

3.29 Measures Taken by Households to Keep Cooked Food Safe in NGP and Non-NGP Villages: MP 99

3.30 Household Level Personal Hygiene Practices-Taking Bath in NGP and Non-NGP Villages: MP 99

3.31 Respondents Awareness about the Existence of VWSC and VHSC in their Villages: MP 100

3.32 Distribution of Households Treating Water Before Consumption and Methods of Making Water Safer for Children's Consumption: MP 102
3.33 Details of In-house Toilet Facility and Type of Toilets Used for Children in NGP and Non-NGP Villages: MP 103
3.34 Distribution of Children Taking Help for Toilet Use in NGP and Non-NGP Villages: MP 104
3.35 Awareness about Toilet Use and Children's Vulnerability to Possible Diseases in NGP and Non-NGP Villages: MP 104
3.36 Distribution of Respondents Reporting Frequency of Bath Given to Children in NGP and Non-NGP Villages: MP 105
3.37 A Distribution of Children with regard to Washing their Hands and Type of Hand Wash Used Before Taking Food in NGP and Non-NGP Villages: MP 106
3.38 Distribution of Households by Immunisation Service Received by Children in NGP and Non-NGP Villages: MP 107
3.39 Number of Children Reported Suffering from Illnesses in NGP Villages: MP. 109
3.40 Number of Children Reported Suffering from Illnesses in Non-NGP Areas: MP 110
3.41 Illness Episodes Encountered, Treatment Sought and Expenditures Incurred by Households in NGP Areas: MP 111
3.42 Illness Episodes Encountered, Treatment Sought and Expenditures Incurred by Households Non-NGP Areas: MP 112
4.1 Demographic Profile of Odisha 127
4.2 Situation of Drinking Water and Water and Sanitation in Odisha. 131

4.3 Gender and Literacy Profiles of the Respondents in NGP and Non-NGP Villages: Odisha. 136
4.4 Profile of Households by Type of House in NGP and Non-NGP Villages: Odisha 137
4.5 Distribution of the Sample Households by Type of Toilet in NGP and Non-NGP Villages: Odisha . . . 137
4.6 Respondents Perceptions Related to the Use of Toilets, by the level of Education in NGP and Non-NGP Villages: Odisha 138
4.7 Details of Hand Wash Habits after Defecation in NGP and Non-NGP Villages: Odisha. 139
4.8 Water Availability and Usage (by Source) in NGP and Non-NGP Villages: Odisha. 139
4.9 Perceptions of the Respondents Regarding the Quality of Water in NGP and Non-NGP Villages: Odisha 140
4.10 Details of Adequacy/Inadequacy of Water Availability for Daily Needs: Odisha 141
4.11 Perceptions of the Respondents Regarding the Adequacy of Water Supply During the Past One Year: Odisha 141
4.12 Particulars of Season-Wise Water Insufficiency (in terms of Intensity): Odisha 142
4.13 Information on the Storage Capacity (Drinking Water) of Water Tanks/ Drums/Vessels in the Sample Households in NGP and Non-NGP Villages: Odisha. 142
4.14 Information on Practices of Drawing Drinking Water from Containers: Odisha 143
4.15 Distribution of Households by Frequency of Cleaning Water Containers Used for Drinking Water in NGP and Non-NGP Villages: Odisha. 144
4.16 Information on Treatment of Drinking Water before Consumption, by Level of Education of the Respondents: Odisha. 145

4.17 Details of Measures Taken for Treating Water before Consumption by the Sample Households in NGP and Non-NGP Villages: Odisha. 146

4.18 Information on the Major Problems Related to Water Supply in the Sample in NGP and Non-NGP Villages: Odisha 147

4.19 Distribution of Respondents Reporting Major Problems Regarding Water Supply in NGP and Non-NGP Villages: Odisha. 148

4.20 Status of Water Supply in ICDS Centres in NGP and Non-NGP Villages: Odisha. 149

4.21 Respondents Awareness Regarding the Advantages of Using In-House Latrine/Toilet/Lavatory Facility in NGP and Non-NGP Villages: Odisha 150

4.22 Respondents Encounter Problems Regarding the Use of Toilets in NGP and Non-NGP Villages: Odisha 150

4.23 Problems Encountered by the Sample Households Relating to Toilet Use in NGP and Non-NGP Villages: Odisha 151

4.24 Information on the Practices of Disposing of Children Stools Post Defecation in NGP and Non-NGP Villages: Odisha 152

4.25 Information on Hand Wash Facility and Type of Hand Wash Near Toilets: Odisha. 154

4.26 Frequency of Cleaning Toilets in NGP and Non-NGP Villages: Odisha 154

4.27 Treatment Seeking Behaviour of the Households with respect to Dehydration among Children in NGP and Non-NGP Villages: Odisha 155

4.28 Measures Taken by the Households to Keep Cooked Food Safe in NGP and Non-NGP Villages: Odisha 156

4.29 Household Level Personal Hygiene Practices: Frequency of Taking Bath in NGP and Non-NGP Villages: Odisha 158

4.30 Respondents Awareness about the Existence of VWSC and VHSC in their Villages: Odisha 158

4.31 Distribution of Households Treating Water before Consumption and Methods of Making Water Safer for Children's Consumption: Odisha. 160

4.32 Details of In-House Toilet Facility and Type of Toilets Used for Children in NGP and Non-NGP Villages: Odisha 161

4.33 Distribution of Children Taking Help for Toilet Use in NGP and Non-NGP Villages: Odisha 161

4.34 Awareness about Toilet Use and Children's Vulnerability to Possible Diseases in NGP and Non-NGP Villages: Odisha 162

4.35 Distribution of Respondents Reporting Frequency of Bath Given to Children in NGP and Non-NGP Villages: Odisha. 163

4.36 Distribution of Children with regard to Washing their Hands and Type of Hand Wash Used before Taking Food in NGP and Non-NGP Villages: Odisha. 164

4.37 Distribution of Households by Immunisation Service Received by Children in NGP and Non-NGP Villages: Odisha 164

4.38 Number of Children Reported Suffering from Illnesses in NGP Areas: Odisha 166

4.39 Number of Children Reported Suffering from Illnesses in Non-NGP Villages: Odisha 168

4.40 Illness Episodes Encountered, Treatment Sought and Expenditures Incurred by Households in NGP Areas, During Last One Year: Odisha 170

4.41 Illness Episodes Encountered, Treatment Sought and Expenditures Incurred by Households in Non-NGP Villages: Odisha 172
5.1 Demographic Profile of AP 184
5.2 Situation of Drinking Water and Sanitation in AP 185
5.3 Gender and Literacy Profiles of the Respondents in NGP and Non-NGP Villages: AP 193
5.4 Profile of the Households by Type of House in NGP and Non-NGP Villages: AP 194
5.5 Distribution of the Sample Households by Type of Toilet in NGP and Non-NGP Villages: AP 195
5.6 Respondents' Perceptions Related to Non-use of Toilets in NGP and Non-NGP Villages: AP 196
5.7 Respondents' Perceptions Regarding the Use of Toilets, by Level of Education in NGP and Non-NGP Villages: AP 197
5.8 Details of Hand Wash Habit after Defecation in NGP and Non-NGP Villages: AP 198
5.9 Water Availability and Usage (by Source) in NGP and Non-NGP Villages: AP 198
5.10 Perceptions of the Respondents Regarding the Quality of Water in NGP and Non-NGP Villages: AP 200
5.11 Details of Adequacy/Inadequacy of Water Availability for Daily Needs: AP 200
5.12 Perceptions of the Respondents Regarding Adequacy of Water Supply During the Past One Year in NGP and Non-NGP Villages: AP 201
5.13 Particulars of Season-Wise Water Insufficiency (in terms of Intensity): AP 201
5.14 Information on the Storage Capacity (Drinking Water) of Water Tanks/Drums/ Vessels in the Sample Households in NGP and Non-NGP Villages: AP 201

5.15 Information on Practices of Drawing Drinking Water from Containers in NGP and Non-NGP Villages: AP 202
5.16 Distribution of Households by Frequency of Cleaning Water Containers Used for Drinking Water in NGP and Non-NGP Villages: AP 203
5.17 Information on Treatment of Drinking Water Before Consumption, by Level of Education of Respondents in NGP and Non-NGP Villages: AP 204
5.18 Details of Measures Taken for Treating Water before Consumption by the Sample Households in NGP and Non-NGP Villages: AP 205
5.19 Information on the Major Problems Related to Water Supply in the Sample Villages (NGP and Non-NGP): AP 206
5.20 Distribution of the Respondents Reporting Major Problems with Regarding Water Supply in NGP and Non-NGP Villages: AP. 206
5.21 Status of Water Supply in ICDS Centres in NGP and Non-NGP Villages: AP 207
5.22 Respondents' Awareness Regarding the Advantages of Using In-House Latrine/ Toilet/Lavatory Facility in NGP and Non-NGP Villages: AP. 208
5.23 Respondents Encountered Problems by the Households Relating to Toilet Use in NGP and Non-NGP Villages: AP. 208
5.24 Problems Encountered by Households Relating to Toilet use in NGP and Non-NGP Villages: AP 209
5.25 Information on the Practices of Disposing Children's Stools Post Defection in NGP and Non-NGP Villages: AP 209

5.26 Information on Hand Wash Facility and Type of Hand Wash Near Toilets: AP 210
5.27 Frequency of Cleaning Toilets in NGP and Non-NGP Villages: AP 212
5.28 Treatment Seeking Behaviour of the Households with respect to Dehydration Among Children in NGP and Non-NGP Villages: AP 213
5.29 Measures taken by Households to Keep Cooked Food Safe in NGP and Non-NGP Villages: AP 213
5.30 Household Level Personal Hygiene Practices: Taking Bath in NGP and Non-NGP Villages: AP 214
5.31 Respondents Awareness Regarding the Existence of VWSC and VHSC in their Villages: AP 214
5.32 Distribution of Households Treating Water Before Consumption and Methods Making Water Safer for Children's Consumption: AP 216
5.33 Details of In-House Toilet Facility and Type of Toilets Used for Children in NGP and Non-NGP Villages: AP 216
5.34 Distribution Children Taking Help for Toilet Use in NGP and Non-NGP Villages: AP. 217
5.35 Awareness About Toilet Use and Children's Vulnerability to Possible Diseases in NGP and Non-NGP Villages: AP 217
5.36 Distribution of Respondents Reporting Frequency of Bath Given to Children in NGP and Non-NGP Villages: AP 218
5.37 Distribution of Children with regard to Washing their Hands and Type of Wash Used before Taking Food in NGP and Non-NGP Villages: AP 219
5.38 Distribution of Households by Immunisation Services Received by Children in NGP and Non-NGP Villages: AP 220

5.39 Number of Children Reported Suffering from Illness in NGP Villages: AP 223
5.40 Number of Children Reported Suffering from Illness in Non-NGP Villages: AP 224
5.41 Illness Episodes Encountered, Treatment Sought and Expenditures Incurred by Households in NGP Areas: AP 227
5.42 Illness Episodes Encountered, Treatment Sought And Expenditures Incurred by Households Non-NGP Areas: AP 228

FIGURES

1.1 A Broad Framework of the Study 49
2.1 Selection of Villages in Contiguous and Widespread Villages 57
2.2 Location of the Study Areas 58
2.3 Visakhapatnam *Mandal/Tehsil* Map 59
2.4 Madhya Pradesh District Map 60
2.5 Odisha District Map 61
3.1 Stored Water in a Cemented Tank in Nawadpura, MP 69
3.2 Drinking Water Facility in a School in Nawadpura, MP 70
3.3 Anganwadi Centre in the School Campus in Pedwi NGP Village, MP 74
3.4 Condition of School Toilet in NGP Village Nawadpura, MP 77
3.5 Pit Toilet in a Household of NGP Village, Pedwi, MP 77
3.6 Water on Roads in Pedwi, a NGP Village, MP 78

3.7 Waste Water in Nawadpura, MP, a NGP Village 79

4.1 Main Source of Water in Odisha: Overhead Tanks Directly to Connected to Pipelines 129

4.2 Main Source of Water in Odisha: Open Well 132

4.3 Wells within Household Premises in Odisha 135

4.4 Posters on How to Make Drinking Water Safer 147

4.5 *Kutcha* House in Odisha 152

4.6 Household Toilet in a Non-NGP Village of Odisha. 153

4.7 Household Toilet in a NGP Village of Odisha 153

4.8 ICDS Centre in Odisha 156

4.9 Washing Clothes Near a Hand Pump in Odisha 157

4.10 An Open Drainage Canal in a NGP Village in Odisha . . . 157

5.1 Main Source of Water: Over Head Tank in Gompa (Non-NGP) Village: AP 187

5.2 *Pucca* Houses in Turakalapudi Non-NGP Village: AP 192

5.3 Village Health Committee Minutes Register in Gompa, AP, a Non-NGP village 198

5.4 Posters Displayed Containing Information on Medicine for Night Blindness and Hand Wash Practicies in Rajupeta: AP 203

5.5 School Children Collecting Water from Hand Pumps in V.J. Puram: AP 205

5.6 School Having Toilets for Boys and Girls in Rajupeta: AP 210

5.7 Toilet Facility in Rayapurajupet Schools: AP. 211

5.8 In-House Septic Latrines in Gompa, AP, a Non-NGP Village 211

5.9 Health Awareness Meeting Conducted in ICDS Centre at Gompa Village: AP. 218

5.10 Distributing Eggs to Children in ICDS Centre in Gompa Village: AP 219
5.11 Hand Wash Practices in ICDS in Gompa Village: AP 220
5.12 Administering Vaccines to Children in Rajupeta Village; AP 221
5.13 Proverbs Written on Walls about the Importance of Toilet Use in Hazipalli: AP 237
5.14 Every Household has Toilet Facility in Hazipalli (NGP) in AP 237
6.1 Main Sources of Water in Three States 242
6.2 Perceptions of the Respondents Regarding the Quantity of Water Available across the Three States . . . 243
6.3 Quality of Water across the Three States. 244
6.4 Frequency of Cleaning Water Containers 245
6.5 Type of Housing Facility 247
6.6 Sanitation Facility 248
6.7 Frequency of Cleaning Toilets 249
6.8 Health Problems among Children. 254
6.9 Morbidity among Adults 256

Foreword

The extensive literature on poverty in India has provided insights into its underlying processes and brought out the multi-dimensional aspects of poverty. The emphasis is now shifting from income poverty to a view that poverty is an outcome of multiple deprivation including attributes like malnutrition and poor health. Any reduction in income poverty due to economic growth may not, pari passu, eradicate malnutrition and ill health. There is now a greater recognition that lack of access to safe drinking water and prevalence of open defecation are some of the major causes of high incidence of malnutrition and ill health. It is in this context, Government of India launched Water and Sanitation Programme to achieve provision of drinking water and sanitation facilities to entire population.

The authors have carried out a field based study on access to safe drinking water, and availability of toilet facility in the villages of Andhra Pradesh, Madhya Pradesh and Odisha. The book *Water and Sanitation Programmes and Health of the Communities: A Study of Three Indian States—Madhya Pradesh, Odisha and Andhra Pradesh* scrutinises the functioning of Village Health and Sanitation Committee (VHSC) and Village Water and Sanitation Committee (VWSC) created under the chairmanship of President of *Gram Panchayats* for monitoring the Total Sanitation Programme (TSP) and assess its impact on health status of the village community. It brings out the constraints including poor governance of the delivery systems at village level in achieving the full potential impact of water and sanitation programmes. The study also shows substantial interstate variation in the outcomes of the programme.

The treatment of the subject is admirably objective and the assessment of the programme is well supported by statistical data. I hope that the book will be useful to policy makers, NGOs (non-governmental organisations), researchers and general public.

—R. Radhakrishna
Chairman
Centre for Economic and Social Studies (CESS)
Hyderabad

Acknowledgements

Many professionals have helped us in the completion of the study entitled "Water and Sanitation Programmes and Health of the Communities: A Study of Three Indian States—Madhya Pradesh, Odisha and Andhra Pradesh". This study would not have been possible without their cooperation. First and foremost, we are grateful to our Chairman, Prof R. Radhakrishna, for his advice during the initial stages of the study with respect to sampling issues and our Director, Prof S. Galab, for his constant encouragement and guidance at various stages of the study.

Special thanks are due to the knowledge partner and study sponsorer, National Health Systems Resource Centre (NHSRC), Ministry of Health & Family Welfare (MoHFW), Government of India, New Delhi; Dr Sunder Raman, former Executive Director; and Dr Sanjiv Kumar, Executive Director, NHSRC, for entrusting the study to CESS. Particularly we thank Dr K. Shashikala, senior consultant, PHP, NHSRC for her guidance and inputs from the initial stages until the completion of the study. We thank Dr Rajesh Narwal, NHRC, for his encouragement and feedback during the initial stages of the study and Dr Sailesh for his constructive inputs during field work in MP.

A number of institutions have extended their help in carrying out our field work. We thank PHCs and SC staff, Anganwadi teachers, *ayahs* and school teachers in the field areas in three states for their support during our field work.

The authors express sincere thanks to Prof Ch. Balaramulu and Dr A. Venkateswarlu for their help during the preparation of the study report. We thank Dr Tapas Kumar Dalpati of MP Institute of Social Science Research, Ujjain, for his help during data collection in

Madhya Pradesh. The authors wish to thank two referees of the study, Prof P.M. Kulkarani, formerly with Centre for the Study of Regional Development, School of Social Sciences, Jawaharlal Nehru University (JNU), New Delhi and Dr B.V. Sharma, Professor of Anthropology, University of Hyderabad (UoH) for their detailed comments and helpful suggestions which have added quality to the study.

We would like to extend our sincere thanks to Ch. Nagaraju, G. Eedaiah and Bishnu Prasad Mahapatra who actively participated and supervised the field work in Andhra Pradesh, Madhya Pradesh and Odisha respectively. We thank Dr Parupally Anjaneyulu, for drawing maps using GIS. We thank all the 15 field investigators who collected data with devotion, commitment and sincerity. Lastly, we thank CESS supporting staff and library for their help.

—Authors

1

Introduction

Rationale Underlying the Research Study

The development of rural India is very important from the view point of achieving inclusive and equitable growth. The analysis of incidence of poverty across Indian states indicates that poverty is very closely related to social infrastructure.[1] An adequate, need based good social infrastructure and effective government interventions through programmes like national drinking water and sanitation programme may result in positive health outcomes. Among the various flagship programmes of the Government of India, the National Rural Health Mission (NRHM) is a unique initiative oriented towards improving the health situation in all the states. NRHM is primarily aimed at strengthening the health infrastructure, increasing community access to the same and creating awareness regarding health related issues. The programme's special focus is on 18 states identified by the Empowered Action Group (EAG) with weak public health indicators and health infrastructure facilities. Further, the macro level data related to morbidity and mortality profiles of the states in rural and urban India points to the underlying social determinants such as nutrition, water and sanitation, among many others. The strength of NRHM lies in working with line departments dealing with the social determinants of health, particularly water and sanitation, in order to achieve the overall objective of improving the health status and quality of life of rural population with an unequivocal and explicit emphasis on sustainable development measures. With this perspective in view, the NRHM promotes community involvement

1. Refers to community facilities like water and sanitation facilities, health facilities to the individuals, families and groups.

at the local level in association with VHSCs (village health and sanitation committees) with a convergent role for facilitating the provisions of health facilities in rural areas.

In this context, an attempt is made to review the relevant literature to identify issues related to water (especially drinking water) across different regions, besides the national drinking water and sanitation policies, problems encountered in grounding these programmes, impact of clean water and sanitation on the health of people and the importance of hygienic practices.

Water and Sanitation

Ensuring drinking water and sanitation security to the people has emerged as a matter of great concern for the governments the world over in the recent times. IGOs (international governmental organisations), NGOs (non-governmental organisations), policy makers, charitable foundations and the private sector are contributing substantially to the debates on what constitutes water and sanitation security and also, how it can be achieved/realised at the global, national, state, community, household and individual levels.

The studies on water and sanitation (health and hygiene aspects) concerning many countries reveal that the availability of and accessibility to protected drinking water, sanitation, personal hygiene, environment, control of diseases and health of the people are closely linked. Thus, an improvement in the health status of people, to a large extent, depends on the water and sanitation related policies, programmes and their effective implementation in terms of making these services readily accessible to the people.

The World Health Organization (WHO) defines the health status as "a state of complete physical, mental and social well-being and not merely an absence of disease or infirmity." The realisation of this goal depends not only on the availability of nutritional food to the people, but also on the health policies/programmes and water and sanitation programmes. The complex, integrated, overlapping social structures and economic systems are responsible for most health related inequities. These social structures and economic systems include the social environment, physical environment, health services, and structural and societal factors. Social determinants of health are shaped by income distribution, power, and resources across local communities, nations, and the world. They must be addressed in terms of achieving many disease-specific targets, including the health-related Millennium Development Goals (MDGs), and controlling and eliminating epidemics that trends to endanger entire populations. Most priority public health conditions share key social determinants, including determinants of exposure to risks, disease vulnerability, access to care, and the consequences of disease (Chandrasekher 2009).

It is pertinent to mention here that the success of water and sanitation programmes, to some extent, depends on the appropriate institutional, financial and regulatory (monitoring) support, active participation of communities at all levels—policy formulation, implementation, evaluation of programmes and sharing/utilisation of schemes—as well as need based training and technical support or access to professional service providers. Thus, there is a need for undertaking an in-depth study on the issues related to water and sanitation programmes vis-à-vis health outcomes for deeper insights into the complex phenomena.

Review of Literature

In the research process, a review of literature helps the researcher understand better the identification and articulation of relationships among the existing/available literature which help in understanding the strengths and gaps in earlier studies. In this section, an attempt is made to present a review of the selected studies/reports pertaining to the present study. An examination of the literature on the performance of the policies, programmes meant for the provision of drinking water and sanitation services reveals mixed views.

Under health outcomes, an attempt is made to discuss the studies examining the impact of poor drinking water and sanitation facilities on the health of population. The impact of hygienic practices is also discussed.

Health Outcomes

Inadequate sanitation conditions continue to remain a leading cause behind a high diarrheal incidence among children in most of the developing countries. A study by World Bank points out that 15 per cent of all the deaths under five years in low and middle income countries are directly attributable to diarrhoea. With respect to children's health, interventions with social pressure and usage of proper sanitation infrastructure can lead to positive health outcomes (Buttenheim 2008).

A study carried out in Peru finds that diarrhoea explains 16 per cent of stunting (Brown et al. 2013). Lin et al. 2013 have shown that children in rural Bangladesh exposed to worse sanitation are more likely to show signs of Enteropathy and are notably shorter, on an average. A 10 per cent increase in open defecation is associated with about a 0.7 percentage point increase in both stunting and severe stunting (Spears et al. 2013).

Stunting afflicts 65 million children in India under the age of five. This disconnect between wealth and malnutrition is so striking that economists have concluded that economic growth does almost nothing to lessen malnutrition (Harris 2014).

Studies conducted in many countries, reveal that disease control, good health, hygiene and water are intrinsically linked. In Malwai, diarrhoea ranks as the third highest cause of morbidity and fifth highest cause of mortality among children under five years of age, according to outpatient health facility and hospital data. Young and Briscoe (1987) conducted a case-control study to test: 1) the hypothesis that improved environmental sanitation to reduce diarrheal incidence; and 2) the feasibility of using a case-control design to evaluate health impacts of environmental interventions. The study shows that, where changes in both water supply and sanitation condition take place, there is a substantial reduction in diarrheal incidence. This study carried out in Malwai region informs that the use of good quality of water supply and latrine facility results in a 20 per cent reduction in diarrhoea in the warm and rainy seasons. A study conducted in Bangladesh reveals that, increase in the proportion of households in the surrounding *basti* that use improved latrines is associated with improvements in child weight-for-height, an important measure of a short-term nutritional status (Buttenheim 2008). Similarly, the water and sanitation intervention project has had a significant impact on diarrheal morbidity among children less than 5 years of age and also among children aged 2-23 months. It is recognised that a cleaner water supply is a necessary condition for reducing the incidence of diarrhoea. The identification of behavioural pattern (e.g. hand washing) relating to contamination might also give better insights into effective intervention strategies (Henry et al. 1990)

Delays in interventions roll-out caused logistical problems especially for the planning of health outcome follow-up surveys. Latrine coverage at the end of the construct period (55%) remained below the target of 70 per cent, a result that may, however, be in line with many other TSC intervention areas in India (Clasen et al. 2012).

Filtu Water and Sanitation Project (FWSP) (in Filtu Woreda of Somali regional state) has been implemented since 2002 in three phases, 2002-2005, 2006-2008 and 2009-2011. Water is a scarce resource in most parts of Filtu Woreda. Solving water problem is also solving other social and development problems. The provision of water is the central and core programme around which the other programme components of FWSP have revolved. Problems identified include water insecurity, poor hygiene, sanitation situation, crop production and others. Drinking water has been made available by harvesting rainwater into partly underground *birkas*, roof catchment and hand dug wells. Sand filters have been introduced as also the health posts. Piloting agricultural activity is done through water pumps. The evaluation team also observed that water projects (mainly *birkas*) will remain sustainable and that the willingness on the part of beneficiaries to pay for operation and maintenance of the Birkas has been impressive (Bussa et al. 2012).

The MDGs have firmly established the issues of "water, sanitation, and hygiene" on the global agenda (Ray et al. 2010). Public health importance of hand washing has also played its role which in reducing of communicable diseases such as diarrhoea and ARI (acute respiratory infections) has been highlighted in many studies

(Ray et al. 2010). As per a study in respect of Tripura, a majority of the respondents practised hand washing after defecation, but it was concerning to note that in both the areas, hand washing was not practised by a substantial proportion of the respondents in situations such as "after changing babies' nappies and disposing off their faeces", "before preparing food", "immediately after handling raw vegetables", and "after handling pets and domestic animals". The reason might be attributed to the similarity in socio-cultural and linguistic backgrounds of the study areas (Ray et al. 2010). Hand washing was a neglected issue as far as the community and care-providers were concerned (including government). Its enormous importance in preventing and reducing the occurrence of diseases such as diarrhoea, ARI and skin infections, based on the evidence, was not properly understood and thus priority has not been given by the community and government to the same extent as given to other programmes such as immunisation, ORT (oral rehydration therapy), RCH (reproductive and child health), etc (Ray et al. 2010).

A study among primary school children in South Kolkata reveals that the commonest morbidity among boys as well as girls was clinical pallor. The study clearly shows that good personal hygienic practices are linked to better health among school children. History of worm infestation was elicited from 28.9 per cent of the children in the Wardha study (Deb et al. 2010).

One of the eight of MDGs, targets agreed by all world governments was to halve world poverty by 2015 and also to halve the proportion of people without access to sanitation. Water Aid Nigeria in its current strategic plan expects to contribute 2 per cent annually to the achievement of MDG target (Water Aid 2007).

Water and Sanitation

The Government of India (GoI) reports, the supporters/advocates of the present policies applaud the Government's efforts in administering the programmes. The studies point out that prioritising of difficult areas; identifying of watershed units; preparation of plan for recharging of groundwater; convergence of National Rural Drinking Water Programme (NRDWP) with Mahatma

Gandhi National Rural Employment Guarantee Act (MGNREGA) and watershed programmes; strengthening of legal, institutional and regulatory issues; improving governance; providing of incentive funds and preparing a management devolution index for the grass-roots level institutions; priority to district planning coordination; participatory planning and implementation of the schemes; ensuring water quality management; sustainable service delivery system; decentralised governance and building professional capacity of the employees etc., would contribute towards a proper implementation of programmes (Department of Drinking Water and Sanitation 2010).

Despite the launching of several programmes and substantial investments in the rural water supply and sanitation sectors, all through the plan periods, the problem of availability as well as quality in the distribution of drinking water is still rampant in various areas across the country. The studies point out commodification of water; high cost/charges for utilisation of water (Wagner 2012); slippage in rural water supply across the states (Ratna Reddy et al. 2010); poor maintenance of environmental sanitation around water points (Rama Chandrudu et al. 2012); a limited access to toilets in the rural areas (a few individuals or households) and variations across the states, regions, urban and rural areas (Kurian Baby and Ratna Reddy 2012); lack of interest on the part of household members in utilising toilets provided by the Government and giving preference to open defecation system (Snehalatha et al. 2012); inadequate budgetary allocation for the construction of latrines; and cornering-off the ILCS schemes by the relatively better-off sections (Alivelu et al. 2012); lack of awareness among residents regarding the programmes, guidelines, allocation of budget and institutions devised for managing the programmes (Ganesh et al. 2013); lack of interest and cooperation on the part of individuals and institutions towards participation in the campaigning of programmes (Pardeshi et al. 2008); non utilisation or non-practice of hand washing by people due to a poor socio-cultural background (Ray et al. 2010); over exploitation of ground water without adequate recharge leading to drying-up of sources or reduced yield; increase in the population; setting up of new habitations; contamination of drinking water sources due to over use of fertilisers, sewage and industrial effluents into ground water etc. (Department of Drinking Water and Sanitation 2010)—as the major challenges/issues which are collectively/individually hampering the progress of water and sanitation programmes in India.

The Eleventh Plan document identifies key issues facing the sector: the main ones are deteriorating source's sustainability resulting from over-extraction of ground water in large parts due to irrigation demand for agriculture, water quality problems including arsenic and fluoride contamination and bacteriological contamination due to lack of sanitation which kills hundreds of children every day in the form of diarrhoea, and poor operation and maintenance including neglect of replacement and expansion resulting in a rapid deterioration in the quality of water services. The other major challenges are related to inter-sectoral coordination, continuous professional support to GPs (*gram panchayats*)/communities and emerging climate change challenges. Further, the studies on sanitation point to lack of awareness, established age-old practice, non-existence of community latrines and insufficient number of latrines, lack of finance/resources for the construction of latrines etc, as the major reasons for the continuance of the open defecation system (Department of Drinking Water and Sanitation 2010).

To accelerate the progress of sanitation in rural areas, the Government of India has designed a paradigm shift in total sanitation campaign (TSC) which is now called the Nirmal Bharat Abhiyan (NBA), in the XIIth Five Year Plan. The objective of NBA is to achieve a sustainable behavioural change in the provision of sanitary facilities to entire communities in a phased, saturation mode with 'Nirmal Grams' as outcomes. The implementation, therefore, needs to concentrate on lower project entities (block/GP).

Water

Here an attempt is made to discuss the government initiated drinking water schemes/programmes. Starting with the Eleventh Plan, the government's endeavour has been to achieve drinking water security at the household level. As per NRDWP,[2] GoI, there is a paradigm shift from just providing water to village to household level with a convenient accessibility of adequate quality water to almost all in India. At the GP level, there is one standing committee called VWSC[3] (village water supply and sanitation committee) (except for Six Schedule Areas).

Despite the presence of good water policies and programmes, slippage is a serious issue and in respect of rural water supply, it amounts to more than 30 per cent at all India level, which is substantial by any standard. Across the states, the extent of slippage varies widely. Thirteen (13) out of 25 states record slippage at rates higher than national average. Slippage ranges from less than 5 per cent in Karnataka and Goa to more than 60 per cent in Mizoram, Arunachal Pradesh and Bihar (Ratna Reddy et al. 2010).

Although proper policies have been put in place, the actual ground reality is not encouraging which is evident from the study conducted in Nirmal Puraskar villages and non Nirmal Puraskar villages of Andhra Pradesh. According to the study, 78 per cent of Nirmal Gram Puraskar (NGP) villages and 57 per cent of non-NGP villages provide adequate water to their citizens. Thirty-nine (39) per cent of NGP villages and 93 per cent of non-NGP villages do

2. To achieve the objective, the state governments adopt appropriate structures depending on the local hydro-geo-morphological conditions suitable to rural drinking water schemes. According to swajaldhara project, water can be defined as safe when it is free from biological contamination (guinea worm, cholera, typhoid etc.) and within permissible limits of chemical contamination excess fluoride, brackishness, iron, arsenic nitrates etc. as per IS-10500 standard of BIS. Priority for new drinking water schemes is to be given to habitations where none (0%) or part of the population has access to adequate and safe drinking water.

3. Participation of representatives of SCs, STs and other backward classes in VWSC should be a priority (DDWS 2010). National Drinking Water Mission ensure transfer of management and financial responsibilities to PRIs, particularly to village water supply and sanitation committees, prioritisation of water usage by the community, protection of ground water from excessive abstraction and reaching safe drinking water to SC/ST dominant habitations. Women's association can provide a strong framework for community participation in water management, as women generally manage domestic water needs.

not maintain environmental sanitation around water points (Rama Chandrudu 2012).

Statement of the Research Problem

Millions die every year of waterborne (bacteria-contaminated water) and water-washed (insufficient water for washing and personal hygiene) diseases (Prüss-Üstün et al. 2008). Diarrhoea alone, a life-threatening symptom of a number of waterborne diseases such as typhoid, cholera, and bacillary dysentery, kills 1.5 million people every year—most of them being children under the age of five (Prüss-Üstün et al. 2008). Malnourished children are at greater risk. The WHO estimates that 10 per cent of all the diseases that affect people in the world each year are related to water quality and access issues, which could be prevented (Prüss-Üstün et al. 2008).

More than 5 million people die each year of diseases caused by unsafe drinking water, lack of sanitation, and insufficient water for maintaining hygiene. In fact, over 2 million deaths occur each year due to water-related diarrhoea alone. At any given time, almost half of the people in the developing countries suffer from water-related diseases (Gleick 2002). Diarrhoea remains the second leading cause of deaths among children under 5 years of age worldwide, after pneumonia. It is responsible for an estimated 1.7 billion cases of diarrhoea, or on average, 2.9 episodes/child/year, and an estimated 1.87 million deaths among children under 5 years of age. The highest burden of disease is borne by children in the age range of 6-11 months: 4.5 episodes/child/year (Brown et al. 2013).

Studies published in 2009 and 2012 focus on the effects of access to safe water, hand washing facilities, and hygiene education on school-age children. A high incidence of diarrhoea contributes to the mortality of about 1.9 million while new diarrhoea cases to the extent of 4 billion (estimated) annually, especially among children under 5 years old. Diarrheal incidences among children during their first few years of life have been shown to limit their growth by 8 cm besides causing an IQ point reduction as they progress to about 7 or 8 years of age. However, hygiene and sanitation interventions have had a considerable impact in terms of reducing diarrhoea incidences and absenteeism rates among school-age children (Joshi and Amadi 2013).

An editorial published in the *Indian Journal of Public Health* reveals that water supplies improve hand hygiene and reduce not only diarrhoea-related child mortality, but also respiratory infections. Yet another study conducted in Rangoon indicates that hand washing is effective in reducing the diarrhoea-related morbidity. The results emerging from studies conducted in Calcutta (Kolkata) reveal that, only a few use soap and water for hand washing before meals (13%), before serving food (1%), before cooking (1%) and after cleaning the child's faeces (5%), while 89 per cent of the respondents in rural areas consider that diarrhoea and dysentery could be prevented by washing hands, but do not give importance to hand washing as a part of preventing incidences of diarrhoea over other methods like maintaining cleanliness, boiling and purifying water before use (Ray 2009).

Programmes Related to Water and Sanitation

In India, although the provision of RWS (rural water supply) is primarily the responsibility of the respective State governments, the Central government contributes a significant part of the programme funds for this sector. During the Five Year Plans, the Central Government has introduced a variety of policies and programmes to address the issue of drinking water. The first national water supply and sanitation programme was introduced during 1951-1956 as part of the Government's health policy. Subsequently the State government's gradually built up their respective Public Health Engineering Department (PHED) for tackling the problem of RWS and sanitation.

The first major push to RWS came with the Accelerated Rural Water Supply Programme (ARWSP) in the 1970s, with full grants to the State governments for implementing water supply schemes in problematic villages. By March 1981, the coverage of RWS was 30.8 per cent. Following the International Drinking Water Supply & Sanitation Decade (IDWSSD) (1981-1991), the second major push came with the establishment of the National Drinking Water Mission (NDWM).

National Rural Drinking Water Programme (NRDWP)

This programme was launched in April 2009 by the then Department of Drinking Water and Sanitation (presently Ministry of Drinking Water and Sanitation-MoDWS), for assisting states in providing drinking water to the rural population in India. 'Drinking Water Security' refers to providing 'every rural person with adequate safe water for drinking, cooking and other domestic needs on a sustainable basis' (NRDWP Guidelines 2010). The GoI strategic plan for rural drinking water goals are:

1 To ensure that every rural person has enough safe water for drinking, cooking and other domestic needs as well as livestock through out the year including during natural disasters.
2. By 2022, every rural person in the country will have access to 70 lpcd within their household premises or at a horizontal or vertical distance of not more than 50 meters from their household without barriers of social or financial discrimination. This indicates that the Government is committed to according a top priority to ensure water security to all the people. It must, therefore, include measures to address source sustainability, water quality and operation and maintenance.

Objectives of NRDWP

The main objectives of NRDWP are to:

1. Enable all households to have access water and use adequate and safe drinking water.
2. Enable communities to monitor and keep surveillance on their drinking water sources.
3. Provide an enabling support and environment for PRIs (panchayat raj institutions) and local communities towards the management of their own drinking water sources and systems in the villages.

4. Ensure a combined approach with respect to rural water supply and rural sanitation so as to achieve saturation of habitations through both these services.
5. Promote participative planning and implementation of water resource management practices.
6. Move away from over dependence on single source to multiple sources through conjunctive use of surface water, groundwater and rainwater.
7. Incentivise states to handover the management of schemes to PRIs through introducing a MDI (management devolution index) based on specific indicators.

Sanitation

The TSC has been renamed as NBA with the objective to accelerate the sanitation coverage in the rural areas so as to comprehensively cover the rural communities through renewed strategies and saturation approaches. NBA envisages covering the entire communities for saturated outcomes. The main objectives of the sanitation programme are to:

- Bring about an improvement in the general quality of life in the rural areas.
- Accelerate the sanitation coverage in rural areas with a view to achieving the vision of 'Nirmal Bharat' by 2022 with all GPs in the country attaining 'Nirmal' status.
- Motivate communities and PRIs in promoting sustainable sanitation facilities through organising awareness creation and health education programmes/campaigns.
- To cover the remaining schools not covered under SSA (Sarva Shiksha Abhiyan) and Anganwadi centres in the rural areas through providing proper sanitation facilities and undertake proactive promotion of hygiene education and sanitary habits among students.
- Encourage cost-effective and appropriate technologies for promoting ecologically safe and sustainable sanitation practices.

- Develop community based and managed environmental sanitation systems by focusing on solid and liquid waste management for achieving an overall cleanliness in the rural areas.

The funds allocated by TSC towards the programme for the year 2007-08 in Odisha amount to 47 per cent while for 2009-10 it was only 32 per cent. In AP, the same amount to only 29 per cent for 2007-08 and 24 per cent for 2009-10. In 2011, Odisha achieved 54 per cent of its physical target of IHHL (individual household latrines) by spending 28 per cent of funds. The second largest component of TSC relates to the construction of school toilets (Accountability Initiative 2011).

GoI has been promoting sanitation coverage in a campaign mode to ensure better health and quality of life for people in rural India. To add vigour to its implementation, GoI launched an award based incentive scheme for fully sanitised and open-defecation free GPs, blocks, districts and States called NGP in October 2003 and gave away the first awards in 2005 as part of its flagship scheme, TSC. NGP till 2011 was given by MoDWS, GoI, at all levels of PRIs, that is, GP, block *panchayat* and district *panchayat*.

The NGP Yojana, a central government scheme for rewards and incentives, and recognition from President of India, was subsequently introduced in 2003, to increase the sanitation coverage in rural areas (Dobe et al. 2011).

In India, water supply and sanitation, which got added to the national agenda during the First Five-Year Plan period (1951-1956), continue to be an important component of planning even today. GoI and State governments keep investing huge amounts on rural drinking water schemes. However, despite the government efforts, the sector continues to be beset with several problems such as deterioration of ground water sources, competing demands for water resources, poor institutional governance, lack of coordination between the government departments, lack of professional services to manage the water supply system, etc, which act as the major hindrances in the implementation of projects.

Table 1.1

Incentive Pattern under NGP:
A Unique Experiment in Sanitation Coverage in India

Particulars	*Gram Panchayat*					*Block*		*District*	
Population Criteria	*<1000*	*1000-2000*	*2000-5000*	*5000-10000*	*10001 and above*	*Up to 50000*	*50001 and above*	*Up to 1 million*	*Above 1 million*
PRIs (₹ lakhs)	0.5	1.00	2.00	4.00	5.00	10.00	20.00	30.00	50.00
Individuals (₹ lakhs)			0.10			0.20		0.30	
Organizations (₹ lakhs)			0.20			0.35		0.50	

Source: NGP Brochure, 2007.

Table 1.2

State-Wise Break Up of Award Winning (NGP) Gram Panchayats

State	*GPs 2006*	*GPs 2007*	*GPs 2008*	*GPs 2009*	*GPs 2010*	*GPs 2011*
Andhra Pradesh	10	143	663	272	44	142
Madhya Pradesh	1	190	682	639	344	212
Odisha	8	33	94	20	81	48
Total	19	366	1439	931	469	402

Source: Department of Drinking Water Supply, GoI, 2011-12.

In spite of the implementation of several programmes during the past six decades, as per Census 2011, only 30.8 per cent of the rural households obtain drinking water from taps, while 22 per cent of the rural households have to walk more than 500 meters for fetching drinking water. Thus, the major challenge before the Central and State governments, is to improve the accessibility of adequate and quality drinking water and sanitation to the people, especially rural communities and thereby, the health conditions of various communities. Access to clean water is a key factor in improving health as it provides scope for reducing poverty indirectly as well as in achieving sustainable development. Freeing women and young girls from the back-breaking travelling long distances to collect water can contribute to achieving gender equity besides improving economic possibilities for families, as women will have more time for income-generating activities, while young girls can attend school (Watkins et al. 2006). Improved health resulting from contamination-free water promises not only a better quality life for people, but also eases pressure on health care systems, and can drastically reduce the number of work days lost due to ill health.

Almost one tenth of the global burden of diseases can be attributed to water sanitation and hygiene. In India, 65 per cent of its rural areas do not enjoy basic sanitation facilities. According to Census 2011, around 53.1 per cent of the people in the country still do not have access to sanitation facilities. A sustainable behavioural change in people and prioritisation of certain control strategies are important in terms of improving the sanitation status of the country (Ganesh et al. 2013). However, there exist variations across different states. A National Sample Survey Organization (NSSO) (2008-09) study shows that access to toilets in rural areas is as low as 8 per cent in Jharkhand and as high as 93 per cent in Kerala (Kurian Baby and Ratna Reddy 2012). According to Census 2011, India, amongst different states, Jharkhand and Odisha tops the list with as high as 78 per cent of homes having no toilet facilities, while the figure is 76.9 per cent for Bihar, 71.2 per cent for Madhya Pradesh and 50.4 per cent for Andhra Pradesh (Census 2011).

Mean latrine coverage among the Odisha villages was 72 per cent (compared to 10% in comparable villages in the same district where the TSC had not yet been implemented), though three of the villages had registered less than 50 per cent of coverage. Among these households with latrines, more than a third of them (39%) were not being used by any member of the household. Over one third (37%) of the members of households with latrines reported never defecating in their latrines. Less than half (47%) of the members of such households reported using their latrines at all times for defecation. Combined with the 28 per cent of households that did not have latrines, it appears that most of the people in these communities still practice open defecation (Barnard et al. 2013).

The implementing agencies tend to focus mainly on the creation of toilet facilities, ignoring sustenance and ecological safety aspects. According to Strategic Plan 2011-2022, the local government institutions in charge of operating and maintaining the infrastructure are ineffective and also lack the required financial resources to carry out their development-related functions. The impact of poor sanitation and resultant episodes of illness invariably lead to the loss of man days work. The household adult members have to either forego their productive labour, or have to stay at home to take care of sick members of the household. Due to a consistent loss of wages, they might as well find themselves trapped in a vicious cycle of poverty.

The new guidelines of NGP (GoI 2012) follow these principles in that 75 per cent of the award money is kept as a fixed deposit in the respective GP's name which is then released only after two years of maintaining the ODF (open defecation-free) status (GoI 2012). However, the financial incentive in itself does not seem to be serving as a prime motivational factor and hence there should be an adequate staggered recognition for sustaining the ODF status. The sustainability should be monitored by employing water quality and health outcomes rather than merely going about constructing toilets for all the households (Kondepati 2013). Most of the slum residents lacking in hygiene with diseases believe that most of the illnesses like fever that affect people, especially children, are caused by flies and mosquitoes which flourish in the unhygienic, dirty conditions surrounding the slums (Suresh Reddy and Snehalatha 2011).

The lack of adequate in-house sanitation can lead to significant economic losses for the country. As per a recent study carried out by Water and Sanitation Programme (WSP), if the economic losses linked to poor sanitation are monetised, the results are staggering. The adverse economic impacts of inadequate sanitation in India due to a poor coverage child mortality etc amount to ₹2.4 lakh crore (US $53.8 billion), or ₹2,180 (US $48) per person as of 2006. This works out to be 6.4 per cent of GDP (gross domestic product) (WSP, Economics of Sanitation Initiative 2010). While the country has come a long way since then and all these indices have improved to a great extent, the linkage between an inadequate sanitation coverage and economic loss is of extreme significance (Department of Drinking Water and Sanitation 2010).

Major Policy Initiatives with respect to Water, Sanitation and Health

The Union Cabinet vide its decision dated 1 May 2013 has approved the launch of National Urban Health Mission (NUHM) as a submission of an over-arching National Health Mission (NHM), with NRHM being the other submission of NHM. Process and outcome indicators are to be developed to reflect equity, quality, efficiency and responsiveness. Targets for communicable and non-communicable diseases will be set at the state level based on the local epidemiological patterns, while taking into account the funding available for each of these conditions. The endeavour towards ensuring the achievement of those indicators is:

1. Reduce IMR to 25/1000 live births.
2. Prevent and reduce mortality & morbidity due to communicable, non-communicable and emerging diseases.
3. Ensure annual Malaria Incidence to be <1/1000.
4. Ensure less than 1 per cent microfilaria prevalence in all districts.
5. Eliminate Kala-azar by 2015; achieve <1 case per 10000 population in all blocks.

The NRHM is an initiative undertaken by the GoI to address the health needs of underserved rural and urban areas. The thrust of the NRHM is on establishing a fully functional, community owned, decentralised health delivery system with inter-sectoral convergence at all levels to ensure a simultaneous action on a wide range of determinants of health such as water, sanitation, education, nutrition, social and gender equity.

NRHM has provided community health volunteers called ASHAs (accredited social health activists) for establishing a link between the community and health system. Their activities include:

- To improve the health care system delivery in the rural areas.
- To supply adequate quantities of essential quality drugs and equipment to PHCs (primary health centres) on a regular basis.

- To envisage a major shift in the governance of public health by giving a leadership role to the PRIs in all matters at the district and sub district levels.
- District health plan would be a reflection of synergy between village health plans, state and national priorities of health, water supply, sanitation and nutrition.
- Visualise the provision of decentralised healthcare at the grass roots level and for this involvement of PRI is considered to be important. An institutional arrangement for constituting VHSCs under the headship of GP is considered important by involving elected GP members of VHSCs for monitoring and implementing of health services at the village level and also for improving the health facility with the slogan 'people health in their hands'.
- Funds to the extent of 26.14 per cent i.e. ₹1811.74 crore have been released under NRHM outlay.
- Untied fund of ₹10,000 to SHGs.
- Release of funds for upgradation of 2 CHCs (community health centres) per district to IPH Standards.

VHSCs will be responsible for the village health plans. Accredited Social Health Activist (ASHA), Aanganwadi Sevika, *panchayat* representative, SHG leader, CBOs (community based organisations) like Parents Teachers Associations (PAT), Mother Teacher Association (MTA), secretary, and local CBO representative will be key persons responsible for the household survey, the Village Health Register and the Village Health Plan. The GP Level Health Plans, comprising a group of villages in many states and a single village in a few, will be worked on at the sub health centre level. The GP Pradhan, Auxiliary Nurse Midwife (ANM), MPW (multipurpose worker), a few VHSC representatives will be responsible for designing the GP Health Plan. They will also be responsible for overview and support for the household survey, preparation of Village Health Registers and Village Health Plans—the GP/SHGs (self help groups) level would also organise activities like health camps to facilitate the planning process (NRHM Framework, 2005-2012).

Sanitation

Pandit Jawarharlal Nehru, the first prime minister of India once said in the early 1950s that "the day every one of us gets a toilet to use, I shall know that our country has reached the pinnacle of progress." Lack of sanitation and the failure to promote hygiene and behavioural change are responsible for the transmission of diarrhoea, sistosomiasis, cholera, typhoid and other infectious diseases affecting millions of human beings. It is estimated that one gram of human excreta contains 10,000,000 viruses, 1,000,000 bacteria, 1,000 parasite cysts and 100 parasite eggs. The sanitation facilities are extremely poor and require lot of improvement for controlling the prevailing waterborne diseases like gastro-enteritis, malaria, diarrhoea, cholera, typhoid, infections, hepatitis and many other diseases caused by various types of viruses/bacteria in the body. Field observations and studies suggest that the sanitation problem is more severe in rural areas where a mere three per cent of the population has access to sanitary toilets (Sharma 2007).

Poor sanitation is also one of the world's leading causes for the spread of diseases and infant mortality. Sanitation refers to the provision of facilities and services for the safe disposal of human waste. Many people lack access to sanitation facilities. Rural sanitation, in particular, came into focus in India during the World Water Decade of the 1980s. The Central Rural Sanitation Programme (CRSP) was started in 1986 to provide sanitation facilities in rural areas. It was a supply driven, infrastructure oriented programme that relied heavily on high levels of subsidies for latrine construction. This approach was subjected to criticism as the sanitation coverage progressed very slowly between 1990 to 2000. The Government has changed the strategy and adopted "Demand driven and community-led approach" replacing the high subsidy approach and launched the TSC approach in 1999. It is a comprehensive programme aimed at ensuring sanitation facilities in rural areas with a broader goal of eradicating the practice of open defecation. To add vigour to the TSC, in October 2003, Government of India initiated an incentive based scheme called 'Nirmal Gram Puraskar'(NGP). NGP is given to those 'open defecation free' Nirmal GPs, Blocks, and Districts which have become fully sanitised. The incentive provision is for PRIs as well as individuals and organizations as the drivers of sanitation movement. The 12th Plan goal is to create an open-defecation free country by 2017 through providing adequate number of community and sanitary facilities.

Sanitation Facilities: Related Experiences in Different States

The main objective of the TSC[4] in India is to accelerate the sanitation coverage in rural areas through facilitating access to toilets for all by 2012. This programme includes "NGPs"[5] which has helped accelerate the pace of implementation of the programme.

Under sanitation programme, VWSC at the village level is a very important institution. However, according to a study carried out in Tamil Nadu, awareness among residents regarding these committees is poor. Out of 15 local residents who were interviewed, only 10 (66.7%) were aware of the number and details of members required for VWSC formation. However, all of them were not aware of the guidelines issued by NRHM with regards to VWSC formation nor did they have any idea about the process of receiving funds for VWSCs (Ganesh et al. 2013).

A study in Maharastra shows how the awards motivated the villagers in improving sanitation in their villages. The sanitation campaign in Yavatmal district of Maharastra was implemented based on the principles and guidelines of TSC with additional incentives and innovations. The interventions were successful in achieving an ODF status for the entire village in a short span of time and was expected to be replicated in other villages. Overall, the campaign aims at mainly achieving an ODF status (Pardeshi et al. 2008). Once the goal of Nirmal Gram status is achieved, it will be challenging to

---

4. For proper solid and liquid waste management, through PRIs, communities and schools,public has to be educated and motivated. TSC encourage use of cost effective and appropriate technologies for ecologically safe and sustainable sanitation for good health. TSC is part of Bharat Nirman and flagship programmes in India. The performance of the project TSC was very satisfactory in 2002-03, therefore the entire central rural sanitation programme has been converted into TSC. Under TSC, the target is to achieve 100 per cent coverage by the end of 11th Five Year Plan (2012).

5. Total sanitation drive include use and maintenance of hygienic latrines, hand wash, water points well managed, waste water disposal in hygienic way and clean public places.

sustain the interest, collaboration, and cooperation on a campaign mode (Pardeshi et al. 2008).

Despite the government's serious efforts, by and large, the results are not satisfactory. According to the survey conducted in 107 villages across nine agro-climatic zones of Andhra Pradesh, only 12 per cent of the households (all the family members) use toilets and only some family members of 22 per cent households use toilets. While all the family members of 66 per cent of the households practise open defecation (Snehalatha et al. 2012).

Sanitation also has socio-economic equity implications. Vulnerable groups (the poor, children, women, the disabled and elderly) tend to suffer the most from the economic impacts of poor sanitation. In fact, diseases associated with poor sanitation have been closely correlated to poverty and infancy which, by themselves, account for about 10 per cent of the global burden of diseases (Minh 2011).

Large variations in the construction activity of latrines have been noticed. For instance, the Government of Andhra Pradesh has sanctioned 2.22 lakh ILCS (integrated low cost sanitation) units in 81 urban centres including Hyderabad city in stage III. The progress however, is low in Rayalaseema (54%) as compared to Coastal Andhra (90%) and Telengana (81%) (Rao et al. 2012). According to another study conducted in Andhra Pradesh, subsidies given for the construction of toilets are not benefitting the deserving and intended beneficiaries (Alivelu et al. 2012).

Relevance of the Study

Although several institutions and scholars have carried out studies as part of evaluating the performance of programmes/schemes meant for providing drinking water and sanitation in terms of households having access to adequate and safe drinking water and sanitation, not many studies have been carried out focusing on the impact of drinking water and sanitation on health conditions of communities, especially in cross-cultural context. The present study is a modest attempt in this direction. The study was undertaken in three states of India (Andhra Pradesh, Madhya Pradesh and Odisha).

Framework/Design of the Study

The study on effects of drinking water and sanitation on health of communities has been designed/conceptualised, keeping in view the fact that water and sanitation and health issues are closely interrelated and that any policy intervention in respect of water and sanitation affects the health status of people. The study aims at analysing the process of implementation of water and sanitation programmes as also identifying the factors that contribute to an effective implementation of policies and also hurdles that hamper progress of programmes. Further, unlike other studies, the present study focuses on the effects of water and sanitation programmes on the health status of people, especially in the rural context. Figure 1.1 describes the framework of the study.

Figure 1.1

A Broad Framework of the Study

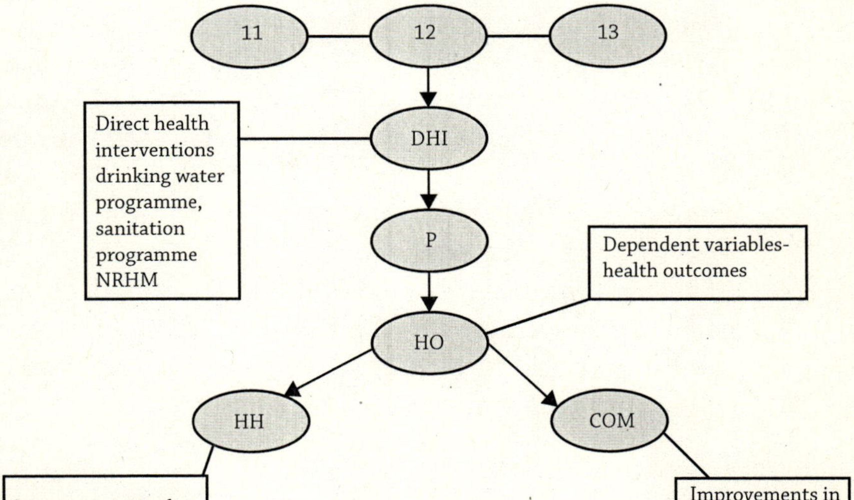

HH - Households

COM - Community

HO - Health outcomes

DHI - Direct health interventions: Drinking Water Programme, Sanitation Programme, NRHM etc

P - Process: The way different institutions interact achieve their goals

11, 12, 13 - Institutions like VWSC, VHSC, SHGs, other CBOs etc at different levels

As shown in Figure 1.1, positive health outcomes can depend on different factors: direct health interventions and other institutional interventions. When health programmes are implemented properly, morbidity levels in the rural areas may come down as also episodes of diarrhoea along with an improvement in nutrition absorption among children etc. However, positive health outcomes largely depend on how various institutions, processes and programmes interact with one another. The study is basically explorative and analytical in nature as it depends heavily on empirical work/field data.

Research Questions

- Whether the provision of sanitation facilities and protected water supply programmes introduced, have paved the way for better health outcomes?
- Whether these programmes have led to any change in the health status of communities in terms of health epidemics (communicable diseases), if so in what way?
- Whether NGP villages with an ODF status, experience a positive/desirable behavioural change in terms of hygiene practices, as compared to non-NGP villages practicing open defecation?
- Whether institutions at the village level have positively impacted the health status of communities?
- What has been the role of the village level institutions/ mechanisms such as VWSCs, VHSCs, CBOs (community based organisations) etc., in the implementation of water and sanitation programmes?

Main Objectives of the Study

- To assess the effects of water and sanitation programmes on the health status of families and health outcomes (including individuals) and to understand the inter-linkages involved in terms of impacting on people's health.

- To understand the role of institutions like VWSCs, VHSCs, SHGs, CBOs etc., in the processes adopted for achieving ODF villages and provision of clean drinking water to all and,
- To understand the role of institutions and processes coming in the way of achieving the ODF status of villages and supply of drinking water.

Hypothesis

- The presence of institutions and their interaction with PRIs and processes adopted in ODF villages for providing safe drinking water and sanitation may be more effective (in terms of fund allocation, utilisation, meetings, decisions taken etc.) as compared to those in non-ODF villages.
- The availability of protected water/sanitation in ODF villages may have a significant impact on the episodes of morbidity at the household level and the incidence of health epidemics at the community level in relation to non-ODF villages.

2

Methodology and Sampling Design

The study employed a multi-stage sampling method for selecting the states, districts, blocks/mandals, villages and households (see Table 2.1). The three states were selected in terms of high-focus and non-high focus, as per the norms stipulated under National Rural Health Mission (NRHM) guidelines (see Figure 2.2). The states of Madhya Pradesh (MP) and Odisha were selected under the high-focus category, while Andhra Pradesh (AP) was selected under the non-high focus category. Following the same criterion/method, i.e. high-focused districts and the number of villages identified/awarded with 'Nirmal Puraskar' (NP), one district each from MP (Dhar) and AP (Visakhapatnam), and two districts from Odisha (Baleswar and Kendujhar) were selected. While selecting the districts, a due weightage was given to those districts where more number of villages have received NP Awards (2011). A Nirmal Gram is an ODF (open defecation free) village where all houses, schools and Anganwadi have access to sanitary toilets and with awareness amongst the community regarding the importance of personal and community hygiene and a clean environment.

Keeping in view the above framework, a due attention was given in to the selection of blocks/*mandals*. From each of the selected districts (4 districts from 3 states), a total of 16 blocks/*mandals* were selected. Six (6) villages from each of the selected districts were selected under NP villages. Similarly, 6 control villages of non-Nirmal Gram Puraskar (NGP) areas from each of the selected districts were also selected. Thus, in all, the study covered three states, four districts, 16 blocks/*mandals*; 36 villages and 720 households (see Table 2.1). While selecting the households in villages, the criterion followed was that there be atleast one child aged between 0 to 5

Table 2.1

Selection and Sample Design

Description of the stage of sampling

S. No	*Sample Unit Description*	*Criteria Followed for Selection*	*Units*	*Remarks*
1	State	Purposive Sampling: 2 states from high-focus category and 1 from non-high focus category.	3	MP and Odisha were selected from high-focus states and AP from non-high focus states.
2	District	Purposive High-Focus District with more number of NP villages.	3	In respect of MP, from among high focus districts with more number of NP villages—Rewa (33 villages), Dhar (22 villages), Sehore (20 villages), Shajapur (14 villages) etc.—Dhar district was selected. In respect of AP, Visakhapatmam district was chosen because it is the only district with a large, number of NGP villages, while other high-focus districts do not have an adequate number of NGP villages for selection.
3	Block/ *Mandal*	Of NP villages spread over more than 3 blocks/*mandals*, a maximum of 3 blocks/ *mandals* were chosen (in order of the *mandals* with highest number of NP villages).	6 blocks in AP, 4 blocks in MP, 6 blocks in Odisha Total=16	In case of selected blocks/*mandals* were contiguous, care was taken to see that the villages were not close to each other. If the sample villages were sparsely spread among many blocks/*mandals*, more than 3 blocks/*mandals* were selected to get the required number of villages.
4	Villages	Among the total NP villages in the sampled district, 6 villages were selected randomly. Six control villages were selected randomly from the sampled blocks/*mandals*.	36 (12 villages from each state) 36 (12 villages from each state)	

5	Households	NP Villages: A total of 60 households were selected randomly from six villages with sanitation facility and similarly, 60 households from the same villages without sanitation facility. Control villages: A total of 60 households were selected randomly from 6 villages with sanitation facility and similarly, 60 households from the same villages without sanitation facility.	720	In the state, from each of the 2 types of households, 120 households were chosen from the sampled villages proportionate to the village population.

Table 2.2

Sampled Villages and Households across the Three States

S. No	*State*	*District*	*NGP Villages*		*Non-NGP Villages*		*Total*
			HHs with Sanitation Facility for Quite Some Time	*HHs with Sanitation Facility Recently*	*HHs with Sanitation Facility for Quite Some Time*	*HHs with Sanitation Facility Recently*	
1	Andhra Pradesh	Visakha patnam	6 villages, 121 HHs		6 villages, 123 HHs		12 villages, 244 HHs
2	Odisha	Baleswar	Baleswar: 6 villages, 121 HHs		Kendujhar: 6 villages, 119 HHs		12 villages, 240 HHs
3	Madhya Pradesh	Dhar	6 villages, 120 HHs		6 villages, 119 HHs		12 villages, 239 HHs
	Total households						723 HHs
	Total villages						36 villages

years. The rationale being the evidenced/established correlation between infant/child health and consumption of safe/clean water and personal and household sanitation, as demonstrated by a number of studies.

The villages have got certain common characteristics like dependence on agriculture, economic inequalities, variations in drinking water and sanitation facilities, and services and access to different government programmes including water and sanitation programmes.

However, these villages/households[1] are diversified in certain respects such as size, location, irrigation facilities, sources and quality of drinking water, sanitation, physical infrastructure facilities and human resources etc. These factors help us capture the ground realities and also the impact of water and sanitation programmes on the health status of the villagers.

Rationale Underlying the Sample Frame

In the present study, it is not intended to give estimates. The nature of the study necessitates an adequate sample to capture variations in behavioural change. Keeping in view the variation, availability of number of NGP villages, time and resource constraints, it was decided to select 6 NGP villages and 6 non-NGP villages from each study state. It is assumed that non-NGP villages are comparable with NGP villages in the study area.

In respect of AP, from among six NRHM high focus districts, only Visakhapanam district has more than 6 'NGP' awarded villages and, therefore, this district was chosen for the study. Out of 7 villages, 6 villages were selected for the study. Another 6 normal villages were chosen as control villages. On the same lines, in MP, there are many high-focus districts with a high number of NGP villages (Rewa-33, Dhar-22, Sehore-20, Shajapur-14). Dhar district with more number of NGP villages was chosen for the study. Following the same guidelines, 6 NGP villages were chosen randomly. Following the same criteria as

1. There are no previous studies on morbidity in the three study states. Hence the sample is largely based on time and budget.

in the case of AP, 6 control villages were chosen from Dhar district. However, in respect of Odisha, identifying adequate NP villages from any of the high focus districts was found to be difficult. Therefore, Baleswar district was chosen for selecting programme villages and Kendujhar (adjacent to Baleswar district and also a high-focus district) was chosen for selecting control villages. Details of sampled villages and households are shown in Table 2.2.

From among the NP villages spread over more than 3 blocks/ mandals, 6 villages were chosen from three different blocks/*mandals*, two from each block/*mandals*. In case these blocks/*mandals* were found contiguous, care was taken to ensure these were not close to each other (Figure 2.1).

Figure 2.1

Selection of Villages in Contiguous and Widespread Villages

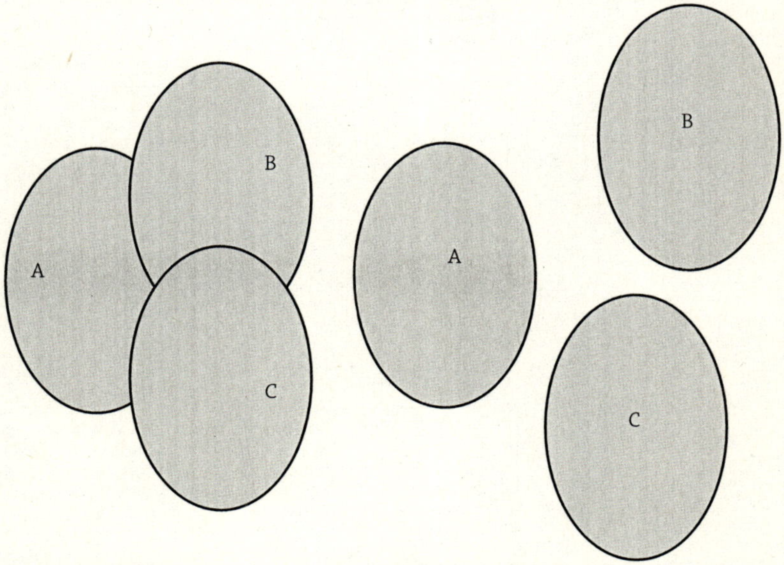

Criterion followed for Infant/Child Presence through House Listing

While selecting households from the villages, care was taken to ensure that there was at least one child aged between 0-5 years in those households on the ground that children are more susceptible to diseases due to lack of sanitation and proper drinking water. For comparison, the age of the children was kept constant. For information on children aged between 0-5 years, house listing (enumeration of all the households in the village) was done in the study villages. In each category of NGP and control, sampled households from each state come to 240 while 720 from all the three states.

Figure 2.2

Location of the Study Areas

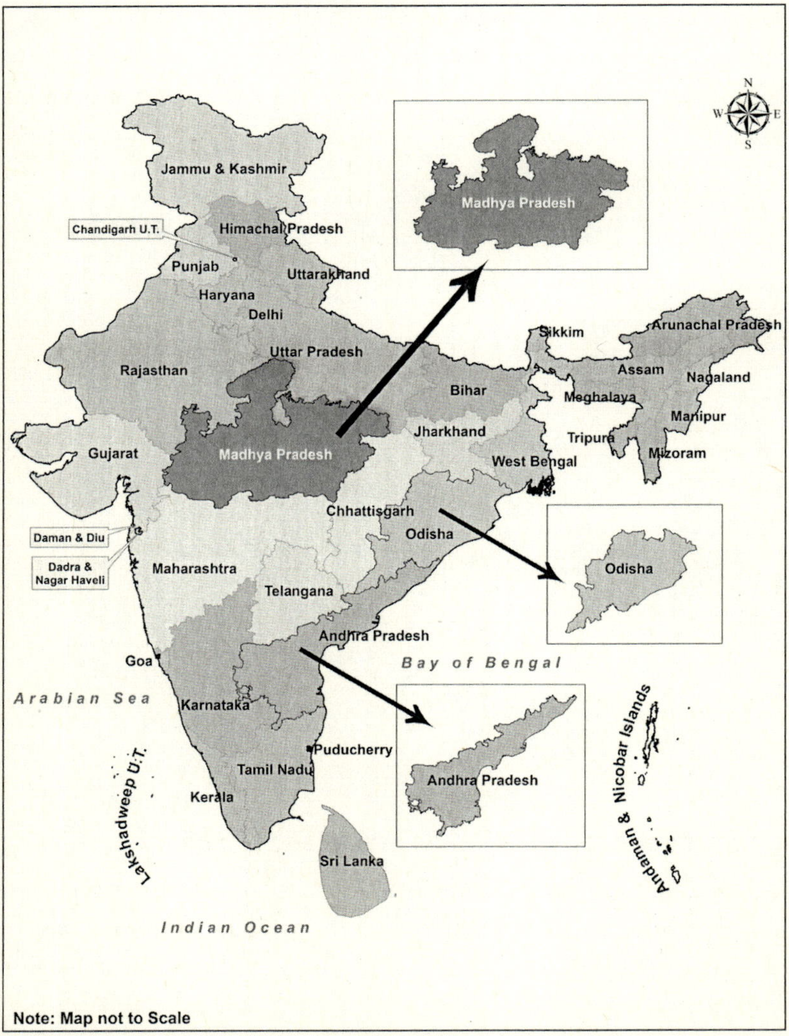

Focus Group Discussions (FGDs)

In each state, 12 FGDs were conducted (6 in NGP and 6 in non-NGP villages). They were conducted with mothers of children aged 0-5 years for collecting information regarding water and sanitation facilities, health and hygiene practices and working of committees in the villages. The pilot study was conducted in November and December 2013 in all the three states. Field work commenced in January 2014 and completed by May 2014.

Andhra Pradesh

The study was conducted in Visakhapatnam district of AP. From the district, six *mandals* were chosen and from each *mandal*, 6 NGP villages and 6 non-NGP villages were chosen for the study (see Table 2.3 and Figure 2.3.

Figure 2.3

Visakhapatnam Mandal/Tehsil Map

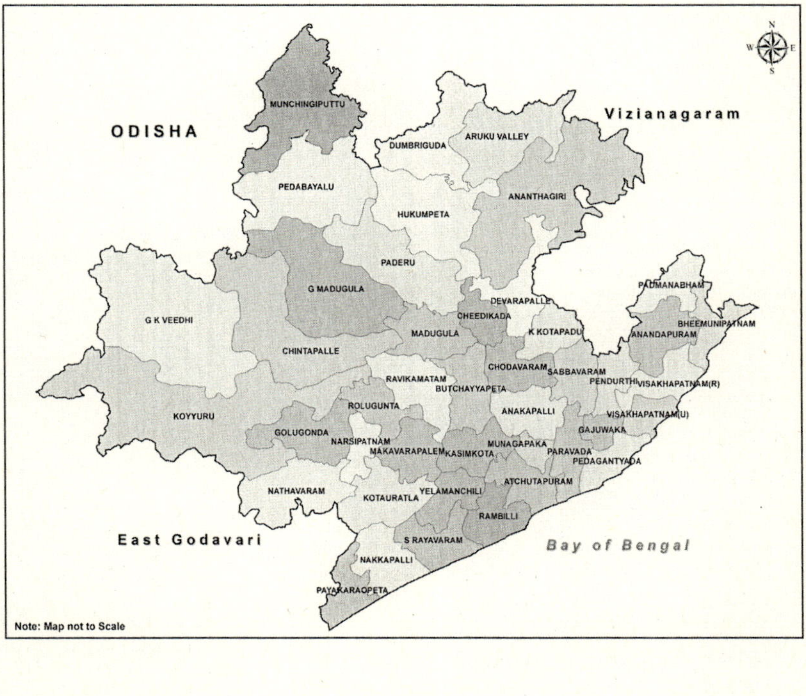

Table 2.3

NGP Villages and Non-NGP Villages in Visakhapatnam District in AP

State	*District*	*Block Name*	*NGP*	*Non-NGP*
Andhra Pradesh	Visakhapatanam	Butchayyapeta	L. Singavaram	Turlapudi
		Choddavarm	Rayapurajupeta	Damunapalli
		Kotauratla	Rajupeta	Ramachandrapalem
		Madugula	V.J. Puram	Gotivad Agraharam
		Ravikamatham	Kavagunta	Gompa
		Yelamanchili	Lakkavaram	Kothali

Madhya Pradesh

Dhar district was chosen for the study and four blocks were chosen from it (see Table 2.4 and Figure 2.4).

Figure 2.4

Madhya Pradesh District Map

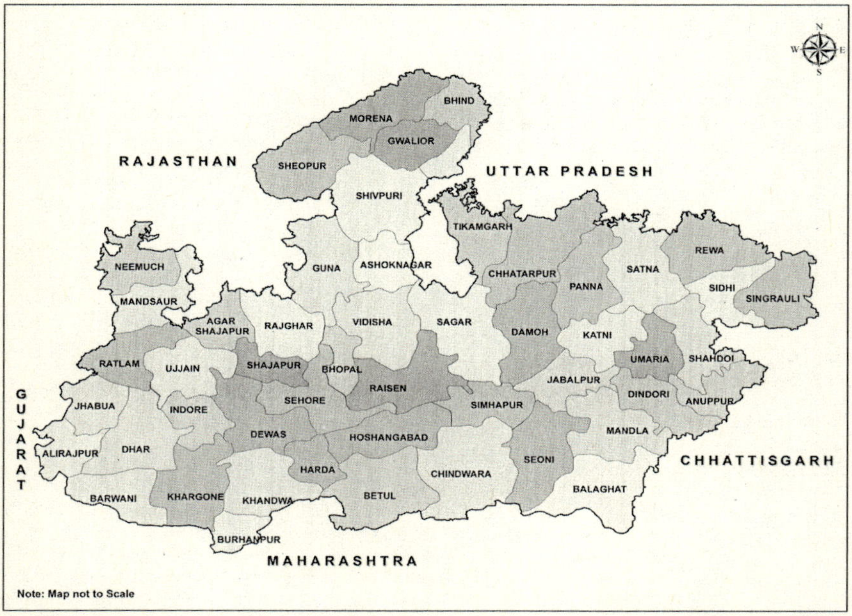

Out of four blocks, 6 NGP villages and 6 non-NGP villages were chosen for our study. The names of the villages are given in Table 2.4.

Table 2.4

NGP Villages and Non-NGP Villages in Dhar District in Madhya Pradesh

State	*District*	*Block Name*	*NGP*	*Non-NGP*
Madhya Pradesh	Dhar District	Badnawar	Nagora	Kachhibaroda
		Badnawar	Pitgara	Hanumantya
		Dharampuri	Morgadhi	Rampura
		Dharampuri	Pedwi	Sala
		Gandhwani	Raipura	Bilda
		Nisarpur	Nawadpura	Amalijhuma

Odisha

In Baleswar district, 5 blocks/mandals were chosen. From these blocks/*mandals*, only NGP villages were chosen, while in Kendujahar district, only Patana block was chosen and from this block, non-NGP villages were selected for the study (see Table 2.5 and Figure 2.5).

Figure 2.5

Odisha District Map

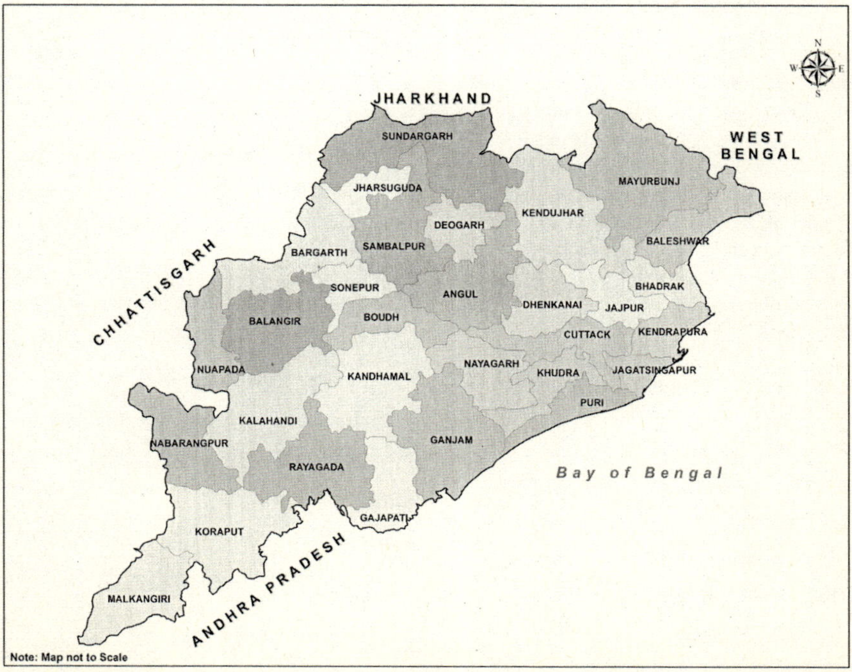

Table 2.5

NGP Villages and Non-NGP Villages in Odisha

S. No	*State*	*District*	*Block/ Mandal*	*NGP*	*District*	*Block/ Mandal*	*Non-NGP villages*
1	Odisha	1. Baleswar (NGP)	Baliapal	Jamakunda	2. Kendu-jahar	Patana	1. Badhikapudi
2			Basta	Naikudi			2. Mushhakhar
3			Jaleswar	Rajpur			3. Rajanagar
4			Nilgiri	K.P. Lakharaj			4. Kendeiposi
5			Remuna	1. Mangalpur			5. Polanghats
				2. Patripal			6. Chinmaliposi

Table 2.6

Study Matrix

Objectives	*Research Questions*	*Variables*	
To assess the impact of water and sanitation programmes on the health status of families including individuals and to understand the factors involved in creating an impact on health (ways in which the communities/families are impacted).	Whether the implementation of provision of protected water supply programmes has led to any positive changes (health behaviour) in the health of families (e.g. reduced illness episodes among family members).	1	Water programmes:
		1.1	Availability of water supply in the villages.
		1.2	Frequency of water supply.
		1.3	Quality of water.
		1.4	Quantity of water.
		2	Hygiene practices
		2.1	Cleaning of water tanks/water containers.
		2.2	Purification of water.
		2.3	Usage of water in the households.
		3	Sanitation programmes:
		3.1	Availability of sanitation facility.
		3.2	Availability of water for sanitary facilities.
		3.3	Frequency of toilet cleaning.
		3.4	Hand wash practices in the households.
		3.5	Usage of toilets.
		4.	Hygiene practices:
		4.1	Measures taken to keep cooked food safer.
		4.2	Measures taken to keep personal hygiene.

To understand the inter link-ages between the provision of safe drinking water and individual sanitation facilities and better health outcomes.	Whether water and sanitation programmes have had a better effect on the health conditions of communities in terms of re-duced epidemics (communicable diseases), if so in what way?	1	Incidence of diarrhoea, malaria etc.
To understand the role of institutions like VWSC, VHSC, SHGs, CBOs etc. and processes adopted for achieving ODF villages and provision of clean drinking water to all.	What is the role of local level institutions such as VWSC, VHSC, SHGs, CBOs etc and processes adopted for achieving ODF villages and provision of clean drinking water?	1	Functioning of local level institutions.
		2	Meetings conducted.
		3	Decisions taken in the meetings for improving water, sanitation facilities.
		4	Funds availability.
		5	Extent of utilisation of funds.
		6	Maintenance of water and sanitation records.
		7	Maintenance of cleaning of water tanks.
		8	Deadlocks in utilisation of funds.
To understand the role of institutions and processes that hinders the achieving of the ODF status of villages and supply of drinking water.	What is the role of institutions and processes that hinder the achieving of the ODF status of villages and supply of drinking water?	1	Poor water and sanitation facilities.
		2	Poor functioning of institutions in the villages.
		3	Deadlocks in the committees.

Pilot Study

After draft schedules were prepared, pilot tests were carried out in the 3 states of AP, MP and Odisha. In AP, a pilot test was carried out in Hazipally, a NGP village in Mahabubnagar district on 27 November 2013. Similarly, a pilot test was conducted in Sadalpur village (NGP) in Dhar district of MP on 14 December 2013. In Odisha also, a pilot test was carried out in December 2013. On the completion of pilot test in the 3 states, the study tools were revised based on the inputs received from the field in terms of rephrasing, changing the order of questions, additions and deletions and so on.

Limitations

- Although NGP villages identified were supposed to be fully equipped with toilet facilities both at the household and school levels in the village, we found during the survey, that the situation in the study villages was different.
- In the study villages of the three states, even though VHSCs (village health and sanitation committee) were present, VWSCs (village water and sanitation committee) had not been formed with respect to water and sanitation programmes for studying the impact of these programmes on the health status of communities every where. It is important to note that measuring behavioural change or health outcomes or health status is possible only when baseline information is available. This study could not access this information and hence, a major limitation of the study.
- Collection of information pertaining to health parameters at the block/*mandal* level was extremely difficult, tedious and time consuming as persons concerned were not available. Whatever meagre data was available, we found it too general in nature, not useful to our study objectives.
- In the study areas, though there are SHGs (self-help groups)/ CBOs (community based organisation)—village education committees, it is clear from the FGDs that they are not actively taking up water and sanitation issues.
- In the study, we have covered VHSCs and VWSCs wherever present.
- The study was intended to collect information on the quality of water, mainly based on the perceptions of the household members. However, it was not possible to assess the quality of water in those areas in the absence of respective institutions.
- The study focus was more on programmes than on policy.

- Project budget, spread of study in the three states and time were also some of the main constraints faced by the study.

Table 2.7

Research Tools Developed and Used for the Study

Tools	*Assessments*
House listing	A household listing schedule was developed for gathering information from each household in the sample villages on sources of drinking water, sanitation facility and also about children between 0-5 years.
Household schedule	A household schedule was developed for gathering information from the head of the family/mother of the child in the household to know about water, sanitation facilities and health conditions; Health seeking behaviour of households; perceptions of household members regarding the functioning of village level committees related to water and sanitation.
Child schedule	A child schedule was developed for collecting information from the mother of the child in the household to know about the health and hygienic conditions; Health seeking behaviour of children aged 0-5 years.
Village schedule	A village schedule was used for collecting information about water, sanitation and health facilities, working of committees with respect to water, health and sanitation in the village from the village head/ *Sarpanch*/VWSC member/VVHSC member/School teacher/ Integrated Child Development Services (ICDS) worker/ASHA/SHG member/NGO member.
Observation tool	This tool was used for assessing the actual situation in the village including water & sanitation practices in schools.
FGDs	FGD guidelines were developed for collecting information from mothers of children aged 0-5 years regarding water and sanitation facilities and health and hygiene practices and the working of committees in the village.

3

Water and Sanitation Programmes and their Effects on the Health Status of Communities in Madhya Pradesh

Introduction

Madhya Pradesh (MP), a central Indian state, is the second largest state in the country in terms of area. Like many other states in the country, it has a rich history and heritage. Bhopal is the capital city of MP; while Ujjain city has a great importance due to its reputation as a major trade centre. The state comprises 50 districts in all.

A Basic Demographic Profile of MP

As per Census 2011, MP has a population of about 7.27 crore, as against 6.03 crore as per 2001 Census. The actual total population of MP, as per 2011 Census, is 72,626,809 of which males account for 37,612,306 and female for 35,014,503 respectively. As per 2001 census, the total population was 60,348,023 of which males numbered 31,443,652, while females are 28,904,371.

The population growth over the decade accounts 20.35 per cent, while for the previous decade to 24.34 per cent. The population of MP constitutes 6 per cent of the total population of India as per 2011 Census, whereas in 2001, it was 5.87 per cent of the total population.

Literacy Rate 2011

The overall literacy rate in MP has seen an upward trend at 69.32 per cent as per 2011 Census. Of that, male literacy stands at 78.73 per cent, while female literacy at 54.49 per cent. As against this in 2001, the overall literacy rate in MP stood at 63.74 per cent of which males and females accounted for 75.35 per cent and 54.61 per cent respectively.

In actual numbers, the total literates in MP stand at 42,851,169 of which males number 25,174,328 and females 17,676,841.

Sex Ratio

Sex ratio in MP is 931/1000 males, which is below the national average of 940, as per 2011 Census. As against this, in 2001 Census, the sex ratio was 920 per 1000 males.

Table 3.1

Demographic Profile of Madhya Pradesh

	2001			*2011*		
	Male	*Female*	*Total*	*Male*	*Female*	*Total*
Population	31,443,652	28,904,371	60,348,023	37,612,306	35,014,503	72,626,809
Literacy Rate (%)	75.35	54.61		78.73	54.49	

Source: Census 2011.

Situation in the Study District and Villages

In Dhar district (selected for the study), 10 per cent of the households are connected with tap water from treated source, while 9.5 per cent of the households are connected with untreated tap water; 16 per cent of the households depend on uncovered well, whereas more than half (53.3%) of the households depend on hand pumps; 6 per cent of the households depend on tube wells/bore hole and 2 per cent of the households depend on rivers/canals (Census 2011).

In Badnawar block, 9.3 per cent, 10 per cent and 8 per cent of the households are connected with tap water from treated source, untreated tap water and uncovered wells, respectively. More than half (55.2%) of the households are dependent on hand pumps, and another 14 per cent on tube wells/bore holes and 1.8 per cent on rivers/canals (Census 2011).

In Dharampuri block, 15 per cent of the households are connected with tap water from treated source and 25 per cent of the households are connected with untreated tap water. More than one fifth of the households (21%) depend on uncovered wells, a little more than one third (34%) of the families depend on hand pumps. 2 per cent of the households depend on tube wells/bore holes and 2 per cent on rivers/ canals (Census 2011).

Water and Sanitation

Sources of Drinking Water

As per Census 2011, the share of households connected with tap water from treated source in MP rural areas amounts to 4.7 per cent, while 5.3 per cent HHs (households) are connected with untreated sources; about one fourth (24%) of the households depend on uncovered wells, 58.3 per cent with hand pumps; about 5 per cent depend on tube wells/bore wells. This indicates that a large number of households do not have access to safe drinking water and that it is one of the biggest challenges for the state and local governments (Census 2011).

Figure 3.1

Stored Water in a Cemented Tank in Nawadpura, MP

Source: Field Work.

Table 3.2

Situation of Drinking Water and Sanitation in MP

Drinking Water	*Urban*	*Rural*	*Total*
Within the household premises	2,130,473	1,446,764	3,577,237
Near the household premises	1,157,105	5,664,862	6,821,967
Away from the households	557,654	4,010,739	4,568,393
Total	3,845,232	11,122,365	14,967,597

Source: Census 2011.

Sanitation Facility	*Urban*	*Rural*	*Total*
Having facility	2,854,081	1,459,201	4,313,282
Not having facility	991,151	9,663,164	10,654,315
Total	3,845,232	11,122,365	14,967,597

Source: Census 2011.

Figure 3.2

Drinking Water Facility in a School in Nawadpura, MP

Source: Field Work.

Sanitation

In rural areas of MP, only 13 per cent of the households have access to in-house latrine facility, while the remaining 87 per cent of the households don't have latrine facility. The type of latrine facilities, that the households have within the premises: 63.4 per cent have septic tanks, 6 per cent piped sewer system and another 9.6 per cent use other systems under flush/pour-flush latrines. Under pit latrine, 13.7 per cent have slab/ventilated improved pits; 6 per cent have no slab/open pits. Among the households without latrine facility, 99.5 per cent practise open defecation and the remaining 0.5 per cent depend on public latrines (Census 2011).

In rural areas of Dhar district, only 22.2 per cent of the households have access to in-house latrine facility and 77.8 per cent of the households go without latrine facility. The type of latrine facility within the premises among the households: 83 per cent of the households have access to septic tanks, 5.4 per cent have piped sewer system and another 4 per cent have other systems under flush/pour-flush latrines. Under pit latrine, 5 per cent of the households have access to with-slab/ventilated improved pits; 2.2 per cent have without slab/open pits. Among the households who don't have latrine facility, 99 per cent opt for open defecation and the remaining one per cent depend on public latrines (Census 2011).

In rural areas of Dharampuri block, a little more than one third (34.4%) of the households have latrine facility within the household, while two thirds (65.6%) of the households don't have access to latrine facility. The type of latrine facilities available within the household premises are: 47 per cent of the households have septic tanks; 6 per cent have piped sewer system and another 10.4 per cent have other systems under flush/pour-flush latrines. Under pit latrine, 19.5 per cent of the households have access to with-slab/ventilated improved pits, 16.6 per cent without-slab/open pits. Among the households who do not have latrine facility, 98.4 per cent practice open defecation and the remaining 1.6 per cent depend on public latrines (Census 2011).

In rural areas of Badnawar block, more than one third (38.6%) of the households have in-house latrine facility within the household, while 61.4 per cent of the households are without latrine facility. The type of latrine facilities within the premises among the households are: 64.8 per cent have septic tanks 4.7 per cent piped sewer system and another 6.7 per cent use other systems under flush/pour-flush latrines. Under pit latrine facility, 16 per cent have with-slab/

ventilated improved pits, 6.8 per cent without-slab/open pits. Among the households without latrine facility, 98 per cent use open spaces for defecation and the remaining 2 per cent depend on public latrines (Census 2011).

A Profile of the Study Villages in the District

For a better understanding of the public policies and governance structures, it is necessary to study the environmental factors like the location, socio-economic structure, infrastructural facilities, civil society bodies, etc. as they wield a considerable influence on the performance of the development programmes. The social development policies such as the provision of protected drinking water and sanitation have to operate in the context of geographical, socio-cultural and political configurations of the society or the areas where they are being operated. The interaction between the environment and administration devised for the implementation of policies is very important, especially at the grass-roots level. In this context, an attempt is made to examine some of the social determinants and infrastructural facilities available in the villages selected for the study. As mentioned earlier, the study was conducted in 12 villages (6 Nirmal Gram Puraskar [NGP] and 6 non-NGP villages) in Dhar district of MP state.

Location and Demographic Features of the Study Villages

The villages are situated at a distance ranging between 25 km to 90 km from the district headquarters. The villages are scattered and some of them do not have their own GPs (*gram panchayats*). The panchayats administration which play a crucial role in the implementation of safe drinking water and sanitation programmes are located away from the residential localities/habitations.

Agriculture is the main occupation of the villagers. The size of land holdings is economically unviable and the poor people tend to migrate to cities in search of employment. The major crops grown in the villages include wheat, jowar, cotton, soya and paddy etc.

The villages are connected by approach roads to GPs' headquarters. The State Road Corporation operates buses to the headquarters of the

GPs. However, many people depend upon private transport and autos managed by private individuals for travelling. The people residing in hamlets do not have proper transport facilities, especially during the rainy season. This causes inconvenience to the people in terms of having a close contact with urban centres, where better service facilities, especially medical facilities, are available. The situation of connectivity is relatively worse in non-NGP villages, as compared to NGP villages. Most of the people in non-NGP areas belong to the tribal communities.

The caste groups in these villages fall under four main categories viz, scheduled castes, schedule tribes, other backward classes and forward classes.

Sources of Drinking Water in the Study Villages

The villagers depend on hand pumps and bore wells, pipelines connected to bore wells and uncovered wells for drinking water. Among these, a majority of them make use of hand pumps and open wells for drinking water, especially in non-NGP villages. In NGP villages, people depend on hand pumps, bore wells and protected water supply for consumption of water.

The field investigation reveals that GPs in NGP villages, supply drinking water through tankers. The villagers also collect water from streams, irrigation tanks, which is not safe for direct consumption. They resort to these methods, as they cannot afford to purchase water from private agencies.

Village Water and Sanitation Committees (VWSCs)

As per National Health Mission (NHM)/programme norms, in every village, there shall be a VWSC to guide and monitor water and sanitation services at the grass roots level. In MP, in 6 NGP villages, 5 VHSCs (village health and sanitation committee) and 2 VWSCs have been formed. Whereas, in non-NGP villages, 4 VHSCs have been formed and there are no VWSCs formed in any of non-NGP villages. The field based observations reveal that about one fifth of the committees do not meet regularly, especially in non-NGP villages. The meetings are mostly held in Anganwadi centres and *panchayat* offices. As many of the households are scattered and are away from GP offices, meetings are held elsewhere, according to the authorities concerned. In many instances, minutes of the meetings are not recorded and the issues relating to water supply are discussed orally, especially in non-NGP villages.

The VWSC has maintained that they pay an adequate attention towards creating awareness among the public regarding waterborne diseases and measures being taken to purify water or alternative methods to procure protected water for maintaining children's health. The data reveals that in nearly two thirds of the villages, especially non-NGP villages, officials are not paying an adequate attention towards creating awareness among the public regarding the importance of safe drinking water.

Figure 3.3

Anganwadi Centre in the School Campus in Pedwi NGP Village, MP

Source: Field Work.

It is remarkable to note that in a majority of the villages—both NGP and non-NGP villages, officials are not preparing health plans. In the absence of plans, the resources, money and time of the functionaries and villagers are being wasted. Further, people do not have a ready access to safe drinking water. FGDs (focus group discussions) with the sample respondents reveal that many of the villagers are not aware of the existence/constitution of VHSC and their functioning at the village level. However, the village sarpanch/secretaries concerned maintain that they do distribute posters, hand-outs to the villagers as a part of creating awareness among them regarding the importance of safe drinking water and proper maintenance of sanitation both at the household and community levels.

State of Water and Sanitation Facilities in Government Schools in the Study Villages

Regarding schools in MP, all the six NGP villages have 6 schools and 6 Indian Child Development Services (ICDS) centres and similarly, all the 6 Non-NGP villages also have 6 schools and 6 ICDS centres. At the school level, observations and interviews with key authorities and teachers reveal that, out of 12 schools, 6 schools use hand pumps (50%) and 25 per cent bore wells as the main source of drinking water in both the NGP and non-NGP villages. There is one school each in non-NGP and NGP villages which have tap connection for supply of safe drinking water. During the field survey, it has been observed that the school authorities are unable to provide safe drinking water for children during school hours.

In non-NGP villages, all the 6 schools use pit latrines, whereas, in NGP villages, 4 schools use pit latrines. Piped sewer system and septic type of latrines are used by NGP schools. However, in NGP village schools, there are separate toilets for girls, boys and teachers. The school management engages/hires the services of private individuals for toilet cleaning, but services are not up to the mark, as the management has failed to pay the charges regularly for the service rendered by individuals. The budget allocation for this purpose is very meagre and, many times, the government does not release the amount on time. As a result, the toilets are cleaned once in a fortnight or a month or sometimes, once in 3 months. This in turn, invariably, cause inconvenience to students in the form of unhygienic or unhealthy conditions and as a result, they hesitate to use toilets. In the absence of adequate budget and proper management, toilets in most of the schools have become defunct.

A Profile of Households and Children's Behaviour Related to Water, Sanitation and Hygiene

This section is based on the data gathered with the help of household and child schedules. Here, an attempt is made to examine the perceptions of household members representing NGP and non-NGP villages regarding the utilisation of water and sanitation programmes and their impact on the health status of the rural communities. The present study drew on the programme (villages under NGP and control-group villages under non-NGP) approach for covering both the NGP and non-NGP village households and for juxtaposing the situation with a view to understanding the perceptions of people towards the programmes and their impact on the health of people.

An analysis of the experiences of those who are recipients of the scheme/NGP is intended to provide insights into the operational dynamics of the programmes. This would also bring out the gap between promise and performance besides highlighting the strengths and weaknesses of the programmes. Similarly, an understanding and assessment of the views of non-NGP households would provide a clue to understanding the cause's underlying the non-availability of facilities provided by the government. Here, an attempt is made to analyse the socio-economic background of the respondents, representing NGP and non-NGP villages.

In this section, an attempt is made to discuss water and sanitation issues and the prevalence of waterborne and other illnesses, based on the data collected from households and children (in the age group of below five years).

The gender particulars of the sample households reveal that an overwhelming percentage of them (94.6%) are males, while an insignificant percentage of them are females. A similar situation is prevailing in both the NGP and non-NGP villages (see Table 3.3). The educational status of the respondents unfolds that most of the female respondents are illiterate, even after the launching of several policies (during the past six and half decades) to improve the educational levels (in the hinterland). And an overwhelming percentage of females (84.62%) are illiterate, while a little over one third of males (39.82%) are illiterate. Among the educated, a majority of them are educated up to primary and secondary levels. A negligible percentage of the respondents (4.8%), especially male members, have completed graduation. It is pertinent to note that the educational levels among females are very low.

Figure 3.4

Condition of School Toilet in NGP Village Nawadpura, MP

Source: Field Work.

Figure 3.5

Pit Toilet in a Household of NGP Village, Pedwi, MP

Source: Field Work.

Figure 3.6

Water on Roads in Pedwi, a NGP Village, MP

Source: Field Work.

Table 3.3

Gender and Literacy Profiles of the Respondents in NGP and Non-NGP Villages: MP

Level of Schooling	*NGP*			*Non-NGP*			*All*		
	Male	*Female*	*Total*	*Male*	*Female*	*Total*	*Male*	*Female*	*Total*
Primary	20	0	20	23	1	24	43	1	44
Per cent	17.5	0	16.67	20.5	14.3	20.17	19.03	7.69	18.4
Secondary (6-12)	47	0	47	35	1	36	82	1	83
Per cent	41.2	0	39.17	31.2	14.3	30.25	36.28	7.69	34.7
Graduate	7	0	7	4	0	4	11	0	11
Per cent	6.1	0	5.83	3.6	0	3.36	4.87	0.00	4.6
Illiterate	40	6	46	50	5	55	90	11	101
Per cent	35.1	100	38.33	44.6	71.4	46.22	39.82	84.62	42.3
Total	114	6	120	112	7	119	226	13	239

The aggregate data reveals that there are variations in the educational levels of the respondents in NGP and non-NGP villages. Illiteracy is more among non-NGP villages (46.22%), as compared to NGP villages (38.33%). These trends indicate that the respondents representing NGP villages have been exposed to a conducive environment (in terms of educational levels) in terms of improving the health status of household members considering that the educational levels of households and the use of protected water along side personal hygiene and sanitation have a positive correlation. Thus, the education levels appear to be one of the preconditions for a proper utilisation of water and sanitation programmes and thereby the protection of health of individuals (Table 3.3).

Figure 3.7

Waste Water in Nawadpura, a NGP Village, MP

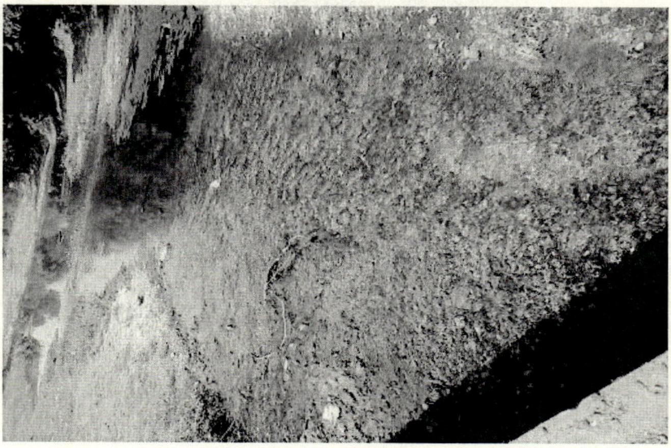

Source: Field Work.

Table 3.4

Profile of Households by Type of House in NGP and Non-NGP Villages: MP

Type of House	*NGP*	*Non-NGP*	*Total*
Pucca	33	23	56
	27.5	19.3	23.4
Semi-*pucca*	28	15	43
	23.3	12.6	18.0
Kutcha	32	32	64
	26.7	26.9	26.8
Hut	27	49	76
	22.5	41.2	31.8
Total	120	119	239

An examination of the availability and type of toilets unfolds that a majority of the respondents use pit toilets (53.1%) and 41.4 per cent of them practise open defecation, indicating the prevailing unhygienic conditions in the residential localities. Open defecation is more visible in non-NGP villages as compared to NGP villages. Only an insignificant percentage of the respondents are making use of flush latrines (0.8%), piped sewerage systems (0.8%) and septic tanks (3.8%). With regards to usage of toilets, non-NGP are going for open defecation and NGP for pit and septic latrine (Table 3.5).

Table 3.5

Distribution of the Sample Households by Type of Toilet in NGP and Non-NGP Villages: MP

Toilet Type	*NGP*	*Non-NGP*	*Total*
Pit latrine	91	36	127
	75.8	30.3	53.1
Flush latrine	2	0	2
	1.7	.0	.8
Piped sewer system	1	1	2
	.8	.8	.8
Septic latrine	2	7	9
	1.7	5.9	3.8
Open defecation	24	75	99
	20.0	63.0	41.4
Total	120	119	239

Note: Chi-value = 54.866, P=O.

An overall picture (both NGP and non-NGP villages together) with respect to the reasons for not using toilets, the response of 20 per cent of the households reveals the non-availability of water either for using toilets or for cleaning. In the case of 8 households (40%), the super structure of toilets does not ensure privacy, while it is 'others' with respect to the remaining 8 households (40%) who reported other maintenance related problems. Out of the total 20, 13 households are from NGP villages who reported the problem in using toilets (Table 3.6).

Table 3.6

Respondent's Perceptions Related to the Non-Use of Toilets in NGP and Non-NGP Villages: MP

Reasons for Not Using Toilets	*NGP*	*Non-NGP*	*Total*
Water is not available for use/cleaning	3	1	4
	23.1	14.3	20.0
Super structure does not ensure privacy	7	1	8
	53.8	14.3	40.0
Others (specify)	3	5	8
	23.1	71.4	40.0
Total	13	7	20

Regarding the reasons for using toilets, 'privacy' is the main reason in more than two thirds of the households (67.1%). A few of the respondents (9.3%) use toilets to maintain better health. Further, the data reveals that a few of the respondents use toilets due to the persuasion of NGOs (non-governmental organisations)/government officials and also to maintain their social status (5.7%).

A comparative picture of NGP and non-NGP villages in utilising toilet facility reveals that a majority of NGP respondents (79.1%) avail themselves of the toilet facility for maintaining privacy and better health, as compared to non-NGP respondents. This indicates that the respondents representing NGP villages are more informed regarding the use of toilet facility as compared to those representing non-NGP villages (See Table 3.7).

The data on the type of housing conditions of the respondents reveals that most of them live either in huts (31.8%) or *kutcha* houses (26.8%), which according to them, are insufficient to accommodate all the family members and are temporary in nature. As the respondents are generally exposed to poor weather conditions during monsoon, winter and summer seasons, they are more likely to be vulnerable to various ailments.

Table 3.7

Respondents' Perceptions Related to the Use of Toilets, by Educational Level in NGP and Non-NGP Villages: MP

Educational Level	*NGP*						*Non-NGP*					*All*					
	For Better Health	*Privacy Status*	*Persuaded by NGOs/ Govt/ Others*	*Social Status*	*Others (specify) Health*	*Total*	*For Better Health*	*Privacy Status*	*Social (specify)*	*Others*	*Total*	*For Better Health*	*Privacy Status*	*Persuaded by NGOs/ Govt/ others*	*Social `Status*	*Others Specify)*	*All*
Primary	5	11	1	1	1	19	2	7	0	3	12	7	18	1	1	4	31
Per cent (%)	26.30	58.00	5.26	5.26	5.26	61.29	16.66	58.33	0.00	25.00	38.70	22.58	58.06	3.22	3.22	12.90	100.00
Secondary (6-12) and Graduation	1	28	0	3	9	41	0	14	0	5	19	1	42	0	3	14	60
Per cent	2.43	68.29	0.00	7.31	22.00	68.33	0.00	73.68	0.00	26.31	31.66	1.66	70.00	0.00	5.00	23.33	100.00
Illiterate	4	27	1	3	1	36	1	7	1	4	13	5	34	1	4	5	49
Per cent	11.11	75.00	2.77	8.33	3.00	73.46	7.69	54.00	7.69	30.76	26.53	10.20	69.38	2.04	8.16	10.20	100.00
Total	10	66	2	7	11	96	3	28	1	12	44	13	94	2	8	23	140

The data unfolds that nearly one third and more than one fourth of the respondents reside in huts and *kutcha* houses respectively, which is indicative of their poor economic status. Besides, only one fourth of the respondents reside in *pucca* houses. The disaggregate data reveals that the respondents representing NGP villages enjoy a better housing accommodation (*pucca* and semi-*pucca*), while Non-NGP respondents have a poor housing accommodation (huts and *kutcha* houses). It is generally observed that people residing in *pucca* and semi-*pucca* houses are relatively better-off and can afford better sanitation facilities, as compared to those residing in huts and *kutcha* houses. This situation tends to affect the health status of people (Table 3.4).

Table 3.8

Details of Hand Wash Habit After Defecation in NGP and Non-NGP Villages: MP

Type of Hand Wash	*NGP*	*Non-NGP*	*Total*
Soap	83	53	136
Per cent	69.2	44.5	56.9
Sand/Ash	14	36	50
Per cent	11.7	30.3	20.9
Plain water	4	1	5
Per cent	3.3	.8	2.1
None of the above	19	29	48
Per cent	15.8	24.4	20.1
Total	120	119	239

The analysis reveals that overall, a great majority of the respondents (both NGP and non-NGP) wash their hands after defecation. The data shows that more than half of the respondents (56.9%) and another one fifth of them (20.9%) wash their hands either with soap or sand/ash respectively, while one fifth of the respondents are neither aware of hand washing nor do they clean their hands after defecation. It is pertinent to point out here that the respondents, who are not following hygienic methods are prone to diseases.

Therefore, it is necessary to educate people and organise intensive campaigns as a part of sensitising people hygiene related practices. This is more so in respect of non-NGP villages where many of the respondents do not wash their hands after defecation as compared to those in NGP villages (see Table 3.8).

Table 3.9

Water Availability and Usage (by Source) in NGP and Non-NGP Villages: MP

Purpose for Using the Occasional Source of Water	*NGP*	*Non-NGP*	*Total*
Domestic use	50	37	87
Per cent	41.7	41.2	36.4
Domestic and toilet use	69	81	150
Per cent	57.5	68.1	62.8
Others	1	1	1
Per cent	.8	.8	.4
Total	120	119	239

One of the criteria for assessing the performance of a project is its accessibility to the people. Regarding the availability of water and its usage, a majority of the respondents use water for domestic and toilet purpose (62.8%) (see Table 3.9).

Generally, even if there is water supply in the villages, the quality of water may not be good and sometimes, it may contain mud or other dust particles. During the survey, the respondents were asked to reveal their perceptions regarding the quality of drinking water. Overall, according to a majority of them, water is pure (54.8%) and occasionally impure going by some others (40.6%). An insignificant percentage of the respondents maintain that water is impure (0.8%). With respect to NGP and non-NGP villages, according to a majority of the respondents (60%), water is pure in NGP villages. In MP, there is difference in mean and also there is a variation in the quality of water. Overall from T-test, we can say that the NGP are more satisfied with the quality of water compared to non-NGP villages (see Table 3.10).

Table 3.10

Perceptions of the Respondents Regarding the Quality of Water in NGP and Non-NGP Villages: MP

Quality of Water	*NGP*	*Non-NGP*	*Total*
Pure	72	59	131
Per cent	60.0	49.6	54.8
Sometimes pure	4	1	5
Per cent	3.3	.8	2.1
Impure	1	1	2
Per cent	.8	.8	.8
Sometimes impure	40	57	97
Per cent	33.3	47.9	40.6
Mostly pure	1	1	2
Per cent	.8	.8	.8
Mostly impure	2	0	2
Per cent	1.7	.0	.8
Total	120	119	239

Note: Chi-value = 8.065, P=0.153.

Table 3.11

Details of Adequacy/Inadequacy of Water Availability for Daily Needs: MP

Quantity of Water	*NGP*	*Non-NGP*	*Total*
Fully sufficient	90	92	182
Per cent	75.0	77.3	76.2
Somewhat sufficient	27	15	42
Per cent	22.5	12.6	17.6
Insufficient	3	11	14
Per cent	2.5	9.2	5.9
Not at all sufficient	0	1	1
Per cent	.0	.8	.4
Total	120	119	239

The availability of adequate quantity of water is also one of the important determinants of health and hygiene conditions of the people. It is also indicative of the performance of service delivery mechanisms of the government or other agencies. The data reveals that more than three fourths of the respondents have access to an adequate volume of water supply to meet their daily needs, while for only an insignificant percentage of the respondents, water supply is inadequate to meet their daily needs (see Table 3.11).

Table 3.12

Perceptions of the Respondents Regarding the Adequacy of Water Supply during the Past One Year in NGP and Non-NGP Villages: MP

Adequacy of Water Supply	*NGP*	*Non-NGP*	*Total*
Yes	52	34	86
Per cent	43.3	28.6	36.0
No	65	82	147
Per cent	54.2	68.9	61.5
Don't know	3	3	6
Per cent	2.5	2.5	2.5
Total	120	119	239

With regard to the adequacy (availability) of drinking water supply during the past one year in NGP villages, more than half of the respondents (54.2%) have maintained that water supply is inadequate, while (43.3%) of the respondents claim that it is adequate (see Table 3.12).

Table 3.13

Particulars of Season-Wise Water Insufficiency (in terms of Intensity): MP

Seasonal Insufficiency of Water	*NGP*	*Non-NGP*	*Total*
Summer	57	66	106
Per cent	83.8	77.6	69.3
Winter	5	8	13
Per cent	7.4	9.4	8.5
Rainy	6	11	17
Per cent	8.8	12.9	11.1
Total	68	85	153

Further, with respect to the seasonal inadequacy of water supply, the field data shows that, on the whole, it is insufficient during all the seasons, but the intensity of the problem is more in the summer season. The perception of NGP and non-NGP village respondents in this respect is, by and large, similar (see Table 3.13).

With respect to the type of drinking water storage facilities, the field data shows that, overall, more than three fourths of the respondents (80%) store water in drums,while some of them store it in pots and tubs (1 tub=20 litres). The respondents, who use pots/tubs/buckets for storing drinking water, represent non-NGP villages. During the field study, it has been observed that the poor households do not have adequate storage facilities and that they face shortage of drinking water. Further it has also been observed that the drinking water supply is not on daily basis but at specified times. As a result, many a time, the poor households face shortage of drinking water. It may be noticed that hygienic practices are not possible because of the shortage of water. This can result in serious health problems (see Table 3.14).

Table 3.14

Information on the Storage Capacity (Drinking Water) of Water Tanks/Drums/Vessels across the Sample Households in NGP and Non-NGP Villages: MP

Storage Capacity of Drinking Water	*NGP*	*Non-NGP*	*Total*
10 Tubs	4	6	10
Per cent	3.3	5.0	4.2
15 Tubs	17	16	33
Per cent	14.2	13.4	13.8
1 Drum	96	93	189
Per cent	80.0	78.2	79.1
10 Buckets	3	4	7
Per cent	2.5	3.4	2.9
Total	120	119	239

Table 3.15

Information on Practices of Drawing Drinking Water from Containers in NGP and Non-NGP Villages: MP

Method of Used for Taking Water from Container	*NGP*	*Non-NGP*	*Total*
Container with a tap attached	1	0	1
Per cent	.8	.0	.4
Tumblers/utensils	62	84	146
Per cent	51.7	70.6	61.1
Tumblers/utensils attached to long handles	57	35	92
Per cent	47.5	29.4	38.5
Total	120	119	239

The data on how drinking water is drawn from containers helps understand the hygiene practices of the respondents. These practices include drawing water from containers with unclean hands leading to waterborne diseases which are very frequent in some seasons. In order to know how the respondents handle drinking water, it was enquired whether: (a) the drinking water container has a tap attached; (b) water is drawn with a glass/utensil using hand; and (c) with a glass/utensil attached to a long handle. To this query, almost all the respondents have mentioned that they always cover pots/drums/buckets, which are meant for drinking water. However, while drawing water from containers for drinking or any other purpose, on the whole, a majority of the respondents (61.1%) use glass tumblers/utensils not attached with long handles. During the field survey, it has been observed that the household members, especially children, draw water from containers without washing their hands leading to ill health of the family members. This situation is more visible in the non-NGP villages as compared to NGP villages. The respondents, who use glass tumblers attached to long handles for drawing drinking water from containers, belong to NGP villages, more than non-NGP villages. Thus, the respondents in NGP villages are more conscious about the consumption of protected drinking water and protecting their health from water related diseases (See Table 3.15).

Table 3.16

Distribution of Households by Frequency of Cleaning Water Containers Used for Drinking Water in NGP and Non-NGP Villages: MP

Cleaning Water Container	*NGP*	*Non-NGP*	*Total*
No	0	0	0
Per cent	0.0	0.0	0.0
Yes	120	119	239
Per cent	100.0	100.0	100.0
Daily	55	54	109
Per cent	45.8	45.4	45.6
Once in 2 Days	54	47	101
Per cent	45.0	39.5	42.3
Once in 5 Days	11	18	29
Per cent	9.2	15.1	12.1
Total	120	119	239

Note: Chi-value 2.918, P=0.232.

The field data shows that all the households clean water containers used for drinking water in both the NGP and non-NGP villages. The data also reveals that a majority of the respondents clean water containers every day or once in 2 days. However, a few of the respondents (12%) wash containers once in five days. With respect to cleaning water containers daily, there is not much difference between NGP and non-NGP villages (See Table 3.16).

Table 3.16 gives information on water treatment before drinking. In NGP villages, out of 92 respondents, who always treat water before drinking, 43.4 per cent are secondary and graduates, 35.9 per cent are illiterates, 20.7 per cent are primary educated. In contrast, only 10 respondents have not at all treated water before drinking, of whom 50 per cent are secondary educated, 30 per cent are illiterates and 10 per cent are primary educated.

In non-NGP villages, out of 82 respondents, who always treat water before drinking, 39.0 per cent are secondary educated, 39 per cent are illiterates, 22 per cent are primary educated. In contrast, only 11 respondents have not at all treated water before drinking, of whom 18.2 per cent are secondary educated, 63.6 per cent are illiterates and 18.2 per cent and primary educated (Table 3.17).

Table 3.17

Information on Treatment of Drinking Water Before Consumption, by Educational Level of the Respondents in NGP and Non-NGP Villages: MP

	NGP					*Non-NGP*					*All*				
Educational Levels	*Always*	*Sometimes*	*Never*	*Not at all Done*	*Total*	*Always*	*Sometimes*	*Never*	*Not at all Done*	*Total*	*Always*	*Sometimes*	*Never*	*Not at all Done*	*Total*
Primary	19	0	0	1	20	18	4	0	2	24	37	4	0	3	44
Per cent	95.00	0	0	5	45.45	75	16.a	0	8.33	54.54	84.09	9.09	0.00	6.81	100.00
Secondary (6-12) and Graduate	40	5	3	6	54	32	4	1	2	40	72	9	5	8	94
Per cent	74.07	9.25	33.33	11	57.44	80.00	10	2.50	5	42.55	76.59	9.57	5.31	8.50	100.00
Illiterate	33	7	3	3	46	32	11	5	7	55	65	18	8	10	101
Per cent	71.70	15.21	7	7	45.54	58	20.00	9.09	12.72	54.45	64.35	17.82	7.92	9.90	100.00
Total	92	12	6	10	120	82	19	7	11	119	174	31	13	21	239

Table 3.18

Details of Measures Taken for Treating Water Before Consumption by Sample Households in NGP and Non-NGP Villages: MP

Measures	*NGP*	*Non-NGP*	*Total*
Boil water	3	1	4
Per cent	2.9	1.0	2.0
Strain water, using a cloth	100	100	200
Per cent	96.2	99.0	97.6
Use water filters	1	0	1
Per cent	1.0	.0	.5
Total	104	101	205

Note: Chi-value = 1.957, P= 0.376.

The field based data shows that almost all of the respondents manually strain (filter) water using a cloth, while negligible percentages of them boil water before drinking. (Table 3.17).

During the field survey, it has been reported that, the non-availability/installation of bore wells in the nearby residential areas, travelling long distances for fetching the water from natural streams/old wells/tanks are the major problems faced by the respondents. Our field observation reveals that the villagers/respondents are not in a position to purchase water from private agencies, paying high amounts. These trends suggest that the government, especially the local governments need to pay more attention to provide safe, cost-free, and adequate quantity of water to the people. Further, according to the villagers, providing safe and adequate quantity of water to people is the responsibility of the local governments (Table 3.19).

There are 45 categories of responses regarding water problems in NGP areas. Major problems reported are: water level goes down during summer (40%); travel long distances to fetch water (37.7%); and go to another place for getting water due to a faulty bore wells (15.5%). In non-NGP areas, the respondents face similar problems like going to far away places for fetching water (44.6%) and going to some other places for fetching water due to faulty bore wells as bore well is not working (21.4%) (Table 3.20).

Table 3.19

Information on the Major Problems Related to Water Supply in the Sample Villages (NGP and Non-NGP): MP

Problems	*NGP*	*Non-NGP*	*Total*
No problems	81	71	152
Per cent	67.5	59.7	63.6
Bore wells not sunk	5	4	9
Per cent	4.2	3.4	3.8
Shortage of water	0	2	2
Per cent	.0	1.7	.8
Long distance involved in fetching water	12	18	30
Per cent	10.0	15.1	12.6
Bore wells is not sunk and shortage of water	5	0	5
Per cent	4.2	.0	2.1
Bore wells not sunk and long distance involved in fetching water	3	8	11
Per cent	2.5	6.7	4.6
Shortage of water and long distance involved in fetching water	3	11	14
Per cent	2.5	9.2	5.9
Others (Specify)	2	1	3
Per cent	1.7	.8	1.3
Bore wells not sunk are absent, shortage of water, long distance involved in fetching water	9	4	13
Per cent	7.5	3.4	5.4
Total	120	119	239

Regarding water supply in ICDS centres (in the case of children attending them), the above tables reveals that, out of a total of 239 respondents from both the NGP and non-NGP villages, 84 have responded ‘yes’, 45 ‘no’ and 110 have no idea about water supply in ICDS centres (Table 3.21).

Table 3.20

Distribution of Respondents Reporting Major Problems Regarding Water Supply in NGP and Non-NGP Villages: MP

Problems	*NGP*	*Non-NGP*	*Total*
Go to another place due to problematic bore wells	7	12	19
Per cent	15.5	21.4	18.8
Get water thrice a week	0	3	3
Per cent	0.0	5.3	2.9
No public taps in the villages	0	2	2
Per cent	0.0	3.5	1.98
No water piped water supply in the villages	1	0	1
Per cent	2.2	0.0	1
Pay money for getting water	1	3	4
Per cent	2.2	5.3	3.9
Water level goes down during summer	18	9	27
Per cent	40	16.0	26.7
Long distance involved in fetching water	17	25	42
Per cent	37.7	44.6	41.5
Unprotected water	1	2	3
Per cent	2.2	3.5	2.9
Total	45	56	101

Note: Multiple answers, percentages exceed 100.

Table 3.21

Status of Water Supply in ICDS Centres in NGP and Non-NGP Villages: MP

Water Supply in Anganwadi Centres	*NGP*	*Non-NGP*	*Total*
Yes	36	48	84
Per cent	30.0	40.3	35.1
No	8	37	45
Per cent	6.7	31.1	18.8
Don't know	76	34	110
Per cent	63.3	28.6	46.0
Total	120	119	239

Table 3.22

Respondents' Awareness Regarding the Advantages of Using In-House Latrine/Toilet/Lavatory Facility in NGP and Non-NGP Villages: MP

Advantages	*NGP*	*Non-NGP*	*Total*
Better health (Hygiene and disease prevention)	21	5	26
	16.9	9.2	14.6
Privacy	86	42	128
Per cent	69.3	77.7	71.9
Ease of use	14	6	20
Per cent	11.2	7.4	11.2
Social status	3	1	4
Per cent	2.4	1.8	2.2
Total	124	54	178

The data on the advantages of using in-house toilet facility shows that a great majority of respondents (69.3% in NGP and 77.7% in non-NGP) use the facility for enjoying privacy; while some respondents (16.9% in NGP and 9.2% in Non-NGP) use the facility for maintaining better health. It is important to note here that most of the respondents prefer to use in-house toilet facility for 'privacy' rather than for maintaining good health. This indicates that the respondents are not fully aware of the advantages of using toilets. Further, as mentioned earlier, considerable proportion of the respondents (40%) does not have access to toilets and depends on open defecation. Therefore, there is a need for organising awareness campaigns to sensitise people to the advantages of using toilets (see Table 3.22).

Table 3.23

Respondents Encounter Problems Regarding the Use of Toilets in NGP and Non-NGP Villages: MP

Problems	*NGP*	*Non-NGP*	*Total*
Yes	46	11	57
Per cent	47.9	25.0	40.7
No	50	33	83
Per cent	52.0	75.0	59.2
Total	96	44	140

The field data shows that many respondents (57 respondents out of 140) encounter problems, while 59.2 per cent of them do not face any problem with regard to the use of toilets (Table 3.23).

In respect of NGP villages, there are 182 multiple responses. Major problems in these villages are: foul smell (19.8%); over-flowing of pits (16.5%); flooding during rainy season (11.5%); mosquitoes/flies menace (16.5%); non-availability of water for cleaning (12.6%); super structure does not ensure privacy (11.5%); and not convenient for children (11%). There are 46 responses in non-NGP areas in the respect. Major problems in non-NGP areas include: water not available for cleaning (17.4%); flies/mosquitoes menace (17.4%); and foul smell (15.2%).

Table 3.24

Problems Encountered by the Households Relating to Toilet Use in NGP and Non-NGP Villages: MP

	Problems	*NGP*	*Non-NGP*	*Total*
1.	Water not available for use/cleaning	23	8	31
	Per cent	12.6	17.4	14.0
2.	Flies/or mosquitoes	30	8	38
		16.5	17.4	16.6
3.	Super structure does not ensure privacy	21	5	26
		11.5	10.9	11.4
4.	Foul smell	36	7	43
		19.8	15.2	18.8
5.	Flooding during rainy seasons	21	6	27
		11.5	13.0	11.8
6.	Difficulties for younger children to use	20	6	26
		11.0	13.0	11.4
7.	Over flowing pits	30	6	36
		16.5	13.0	15.7
8.	Unstable slabs (Fear of falling)	1	0	1
		0.5	0.0	0.4%
	Total	182	46	228

Note: Multiple answers, percentages exceed 100.

The non-availability of an adequate quantity of water for cleaning toilets, an uncontrollable menace of mosquitoes in the household premise and residential localities, lack of privacy, bad smell due to an irregular maintenance of drainage canals, lack of connectivity of toilet pipes with the under drainage system, blockage of toilet pipes during rainy season, unloading or irregular cleaning of toilet pits, and unsafe toilets for usage due to weak walls/structures and others, are the major problems encountered by the respondents in using toilets. Some of these issues are related to the linkage of toilets to the main drainage system in the villages and general environment. The local governments' role, in this context, is of utmost importance in terms of controlling mosquitoes and properly maintaining the drainage system in the residential areas (see Table 3.24).

Table 3.25

Information on the Practices of Disposing of Children's Stools Post Defecation in NGP and Non-NGP Villages: MP

Method of Disposal	*NGP*	*Non-NGP*	*Total*
Leave it where it is	1	1	2
Per cent	.0.8	0.8	0.8
Throw it on the street	71	77	148
Per cent	59.2	64.7	61.9
Throw it in the latrine	40	33	73
Per cent	33.3	27.7	30.5
Others	8	8	16
Per cent	6.7	6.7	6.7
Total	120	119	239

Note: Chi-value = 0.91, P= 0.823.

Regarding practices of disposing of child's stool post defecation, most of the respondents (61.9%) in both the areas throw it on the street, while 30.5 per cent of the respondents throw into the latrine. In respect of NGP areas, 59.2 per cent of the respondents and most of the respondents (64.7%) in non-NGP villages throw it on the streets. Thus in both NGP and non-NGP, most of the respondents dispose childs stool post defecation on the street (Table 3.25).

With regard to the availability of facilities for hand wash near the toilet at home, a great majority of respondents are affirmative (81%) in that they wash their hands with soap in both NGP and Non-NGP villages together, while another 7.3 per cent of them use soap/ash for washing their hands after defecation (see Table 3.26).

Table 3.26

Information on Hand Wash Facility and Type of Hand Wash Near Toilets: MP

Type of Hand Wash	*NGP*	*Non-NGP*	*Total*
Not having wash facility	3	0	3
Per cent	3.1	.0	2.1
Having wash facility	93	44	137
Per cent	96.9	100.0	97.9
Soap	74	37	111
Per cent	79.6	84.1	81.0
Sand/Ash	7	3	10
Per cent	7.5	6.8	7.3
Plain Water	6	3	9
Per cent	6.5	6.8	6.6
Soap and Sand	6	1	7
Per cent	6.5	2.3	5.1
Total	93	44	137

Table 3.27

Frequency of Cleaning Toilets in NGP and Non-NGP Villages: MP

Frequency of Cleaning Toilet	*NGP*	*Non-NGP*	*Non-NGP*
Once or more in a day	36	17	53
Per cent	37.5	38.6	37.9
At least once in a week	11	7	18
Per cent	11.5	15.9	12.9
At least once in a month	11	5	16
Per cent	11.5	11.4	11.4
Others	36	12	48
Per cent	37.5	27.3	34.3
Total	96	44	140

A proper maintenance of toilets is most important for keeping good health. The field data reveals that, overall, only a little over one third (37.9%) of the respondents clean their toilets everyday, and the rest of them, especially the respondents representing non-NGP villages trend to neglect cleaning their toilets. A considerable percentage of the respondents clean their toilets once in a month, which makes it unsafe to use such toilets (Table 3.27).

In this section, illnesses such as diarrhoea, dysentery and worm infections related to unsafe drinking water and unhygienic practices followed in the study villages resulting in health problems for children are discussed. Further, methods followed by the households for treating dehydration of children are also presented here.

Table 3.28

Treatment Seeking Behaviour of the Households with Respect to Dehydration among Children in NGP and Non-NGP Villages: MP

Method Followed in Treating Dehydration	*NGP*	*Non-NGP*	*Total*
Provide ORS	1	0	1
Per cent	.8	.0	.4
Provide only hot water	1	0	1
Per cent	.8	.0	.4
Visit Doctor/clinic	115	110	225
Per cent	95.8	92.4	94.1
Get medicine from pharmacy	1	1	2
Per cent	.8	.8	.8
Approach quacks in the village	1	5	6
Per cent	.8	4.2	2.5
Visit ASHA/ANM	1	3	4
Per cent	.8	2.5	1.7
Total	120	119	239

Majority of the respondents are not aware of information regarding using ORS (oral rehydration solution) whenever children suffer from dehydration. Only one respondent from NGP villages is aware of the usefulness of ORS. Overall, a majority of the respondents

(94.1%) prefer to visit a doctor whenever their children suffer from dehydration. Interestingly, a very few households approach village quacks (2.5%) and Accredited Social Health Activist (ASHA)/Auxiliary Nurse Midwise (ANM) (1.7%) (Table 3.28).

Table 3.29

Measures Taken by Households to Keep Cooked Food Safe in NGP and Non-NGP Villages: MP

Measures Taken	*NGP*	*Non-NGP*	*Total*
Cover cooked food with a lid	119	117	236
Per cent	99.2	98.3	98.7
Do not bother	1	2	3
Per cent	.8	1.7	1.3
Total	120	119	239

Keeping the cooked food safe by covering the same with a lid is a safe way of maintaining hygiene and preventing occurrence of diseases. Interestingly enough, the field data shows that an overwhelming number of the respondents (NGP and non-NGP villages taken together) (98.7%) cover cooked food with a lid, while only a very tiny per cent (1.3%) of the respondents do not keep cooked food properly (Table 3.29).

Table 3.30

Household Level Personal Hygiene Practices: Taking Bath in NGP and Non-NGP Villages: MP

Frequency of Taking Bath	*NGP*	*Non-NGP*	*Total*
Once or more a day	80	70	150
Per cent	66.7	58.8	62.8
At least once in 2 days	36	41	77
Per cent	30.0	34.5	32.2
Once in a week	4	8	12
Per cent	3.3	6.7	5.0
Total	120	119	239

Regular bathing is considered as one of the hygiene practices for keeping good health. The field data shows that, overall, two thirds of the respondents (62.8%) take bath once in a day and another one third of them once in two days. A few of the respondents (5%) take bath once in a week, indicating their poor health consciousness (Table 3.30).

VWSCs and VHSCs in the Study Villages

An attempt is also made to know the respondents' awareness level of the existence and working of VWSC in their respective villages. It is pertinent to note that overall, except for one respondent, all the respondents (99.6%) are not aware of the existence of the local level committee, not to speak of its working. Similarly, almost all the respondents (99.6%) are also not aware of the existence of VHSC and its working. These trends indicate that the government agencies/machinery have failed to create awareness among the public about the existence of functionary/committee. The elite, media and civil society bodies also have failed to notice the existence of the basic committees meant for the implementation of water and sanitation programmes in the rural areas. It also indicates that there is a lack of

Table 3.31

Respondents Awareness about the Existence of VWSC and VHSC in their Villages: MP

Existence of VWSC	*NGP*	*Non-NGP*	*Total*
Aware	1	0	1
Per cent	.8	.0	.4
Not aware	119	119	238
Per cent	99.2	100.0	99.6
Existence of VHSC			
Aware	1	0	1
Per cent	.8	.0	.4
Not aware	119	119	238
Per cent	99.2	100.0	99.6
Total	120	119	239

community participation in the policy formulation, implementation, evaluation and sharing of benefits of the programmers at the local levels (Table 3.31).

An Analysis of Information Based on Child Health and Hygiene

As part of the study, an attempt has also been made to collect information on various dimensions of child health and hygiene. This was carried out by administering an interview schedule for mothers of children aged 0-5 years. The purpose of the study/schedule is to identify health and hygiene practices followed by parents, especially mothers with respect to their children. The schedule covered various aspects such as practices treating drinking water; awareness about diseases/ailments children may be suffering from, due to unsafe drinking water; the availability of toilet facility especially meant for children; diseases/ailments affecting children as a result of open defecation practice or non-use of toilets; personal hygiene practices; improper bathing and washing of hands and material used for washing. The child schedule also contained questions related to the status of immunisation of children and maintenance of immunisation card separately for each child and illness episodes underwent by children during the last one year. The data collected on these items is intended to help us have a better idea not only regarding the health status of our future generations, but also strengths and weaknesses of the policies meant for human development, especially an improvement in the health status of children.

With regard to providing drinking water for children's consumption, overall more than half of the respondents use clean and safe water. In addition to this, breast feeding and providing nutritious food and maintaining a healthy environment in and around the households and residential localities also play a crucial role in keeping children healthy, according to the respondents. Further, the respondents expect the government to give a priority to these issues, instead of leaving it to the households. Hence, going by their perceptions, an effective implementation of ICDS schemes, water and sanitation schemes, food security programme, and environmental protection measures can go a long way in improving the health conditions of children in particular and household members, in general. However, overall, a little over one fifth of the respondents (7%) find themselves unable to treat drinking water before giving it to children.

Lack of awareness regarding child healthcare practices, poor economic conditions, inability to give a special attention to health care, especially treating drinking water before giving to children (due to their pre-occupation with their routine household activities and other activities), lack of availability of adequate and qualitative water in the villages, inability of local governments or agencies to supply treated water at ICDS centres or within the premises of households, etc, are the major reasons for this state of affairs in the rural areas, according to the respondents.

Table 3.32

Distribution of Households Treating Water Before Consumption and Methods of Making Water Safer for Children's Consumption: MP

Treatment of Water	*NGP*	*Non-NGP*	*Total*
Yes	69	73	142
Per cent	57.5	61.3	59.4
No	51	46	97
Per cent	42.5	38.7	40.6
Total	120	119	239
Methods Used for Making Water Safer for Children's Consumption			
Boil water	5	4	9
Per cent	7.2	5.5	6.3
Strain water using a cloth	62	60	122
Per cent	89.9	82.2	85.9
Use water filters	0	1	1
Per cent	.0	1.4	.7
Let water stand and settle	2	8	10
Per cent	2.9	11.0	7.0
Total	69	73	142

The disaggregate level data reveals that the households/respondents, who give clean and safe drinking water to their children, are more in numbers in non-NGP villages than NGP villages. It indicates that the local governments/public agencies' role in providing safe water to people is more important for maintaining the health of children. When it comes to the measures followed for making water more cleaner and safer for drinking, especially for children, overall, a majority of the respondents (85.9%) clean water by straining or filtering using a cloth. Some of them let water stand and settle (7%) while 6.3 per cent of the respondents boil water to make it safer for children's consumption. Overall from the T-test, in MP no significant difference has been observed in treating water for both NGP and non-NGP villages (Table 3.32).

Coming to the availability of in-house toilet facility, especially for the use of children, overall, more than half of the respondents (56.9%) have expressed themselves in affirmative, while the remaining 43.1 per cent do not have access to such in-house facility. The respondents, who do not have such in-house toilet facility number are more in non-NGP villages as compared to NGP villages. Three-fourths of the respondents (77.5%) in NGP and about one third of the respondents (36.1%) in non-NGP areas have access to in-house toilets facility (Table 3.33).

Table 3.33

Details of In-house Toilet Facility and Type of Toilets Used for Children in NGP and Non-NGP Villages: MP

Toilet Facility	*NGP*	*Non-NGP*	*Total*
Yes	93	43	136
Per cent	77.5	36.1	56.9
No	27	76	103
Per cent	22.5	63.9	43.1
Type of Toilets Used for Children In-house			
Toilet within the house	90	43	133
Per cent	75.0	36.1	55.6
Open defecation	24	75	99
Per cent	20.0	63.0	41.4
Others	6	1	7
Per cent	5.05	.8	2.9
Total	120	119	239

Table 3.34

Distribution of Children Taking Help for Toilet Use in NGP and Non-NGP Villages: MP

Children Taking Help for Toilet Use	*NGP*	*Non-NGP*	*Total*
By his/her own	12	9	21
Per cent	10.0	7.6	8.8
With the help of mother/father	106	108	214
Per cent	88.3	90.8	89.5
With the help of elder sibling/s	1	1	2
Per cent	.8	.8	.8
With the help of others	1	1	2
Per cent	.8	.8	.8
Total	120	119	239

The data unfolds that an overwhelming 89.5 per cent of the respondents (mother/father) help their children, especially those below two years, in using toilets, while an insignificant percentage of respondents (elder sibling/s and other members of the household) help children in this regard. However, according to less than 10 per cent of the respondents, children use toilets on their own and do not require others' help as they are grown up (Table 3.34).

Table 3.35

Awareness about Toilet Use and Children's Vulnerability to Possible Diseases in NGP and Non-NGP Villages: MP

Type of Diseases	*NGP*	*Non-NGP*	*Total*
Diarrhoea	5	12	17
Per cent	4.2	10.1	7.1
Cold, cough, fever	0	2	2
Per cent	.0	1.7	.8
Fever, cold	5	3	8
Per cent	4.2	2.5	3.3
Don't know	110	102	212
Per cent	91.7	85.7	88.7
Total	120	119	239

The above table reveals that, overall, a majority of the respondents (88.7%) have no idea about the type of disease children may be vulnerable to, due to open defecation and non-use of toilets on the part of children. Only 7 per cent of the respondents are aware that children may get diarrhoea if they do not use toilets. The level of awareness in this respect seems to be relatively much better in non-NGP villages (Table 3.35).

Table 3.36

Distribution of Respondents Reporting Frequency of Bath Given to Children in NGP and Non-NGP Villages: MP

Frequency of Bath Given to Children	*NGP*	*Non-NGP*	*Total*
Once a day	75	65	140
Per cent	62.5	54.6	58.6
Twice a day	33	35	68
Per cent	27.5	29.4	28.5
Once a week	12	18	30
Per cent	10.0	15.1	12.6
Others (specify)	0	1	1
Per cent	.0	.8	.4
Total	120	119	239

An analysis of the practice of personal hygiene reveals that, overall, more than half of the respondents (58.6%) help their children take bath every day, especially the respondents representing NGP villages (62.5%), while more than one fifth of the respondents (28.5%) give bath to their children twice a day. Some of the children (12.6%) take bath once in a week and they mostly belong to non-NGP villages. Thus, children in non-NGP villages are at a disadvantageous position as compared to NGP villages (Table 3.36).

In both NGP and non-NGP areas together, according to majority of the respondents (90.8%), their children wash their hands before taking food,while as per more than half of the respondents and over one fifth of the respondents, their children use soap and water for washing hands respectively. The respondents representing NGP villages are more conscious about washing hands with soap as compared to non-NGP villages (Table 3.37).

Table 3.37

A Distribution of Children with regard to Washing their Hands and Type of Hand Wash Used Before Taking Food in NGP and Non-NGP Villages: MP

Children Washing their Hands	*NGP*	*Non-NGP*	*Total*
Yes	115	102	217
Per cent	95.8	85.7	90.8
No	2	13	15
Per cent	1.7	10.9	6.3
Don't know	3	4	7
Per cent	2.5	3.4	2.9
Total	120	119	239
Type of Hand Wash Used			
Water	22	37	59
Per cent	19.1	36.3	27.2
Soap	85	51	136
Per cent	73.9	50.0	62.7
Others	0	5	5
Per cent	.0	4.9	2.3
Water soap	8	9	17
Per cent	7.0	8.8	7.8
Total	115	102	217

The data reveals that half of the respondents in NGP villages and in non-NGP villages have got their children immunised for keeping good health, while as per a small percentage of respondents i.e. 1.7 per cent in NGP and 5.9 per cent in non-NGP villages, their wards are yet to receive vaccination to protect their health (Table 3.38).

Out of 185 children, 139 in NGP areas and out of 169 children, 131 in non-NGP areas have been found experiencing various illnesses.

Table 3.38

Distribution of Households by Immunisation Service Received by Children in NGP and Non-NGP Villages: MP

Immunisation	*NGP*	*Non-NGP*	*Total*
Received	69	52	121
Per cent	57.5	43.7	50.6
Not received	10	16	26
Per cent	8.3	13.4	10.9
Partly received	5	9	14
Per cent	4.2	7.6	5.9
Yet to receive immunisation	2	7	9
Per cent	1.7	5.9	3.8
Don't know	34	35	69
Per cent	28.3	29.4	28.9
Total	120	119	239

Children Affected by Illnesses

In non-NGP areas, 25 diarrheal cases and 29 cases in NGP areas have been reported. In non-NGP areas, 5 malaria child cases and in NGP areas only 2 malaria cases have been reported. In non-NGP areas, 8 children and 5 children in NGP areas have been reported affected by pneumonia.

In non-NGP areas, 33 children have been reported suffering from cough, cold and fever together and 30 children in NGP areas. A few cases of jaundice and skin diseases have also been reported in non-NGP areas.

Treatment

In both the areas, a majority of the child patients were taken to private medical practitioners or private hospitals for treatment, while a very few children were taken to SC (sub centres), PHC (public health centres) and DHC (district health centres) for treatment.

In both NGP and non-NGP areas, majority of the child patients of diarrhoea received treatment from private hospitals and a few from PHC/DHC. Three child patients from non-NGP areas and only one patient from NGP areas received treatment for more than a day. In non-NGP areas, more number of children had to seek treatment for cold, cough, fever for more than a day.

A disease-wise and length of treatment analysis shows that the severity of illnesses is more in non-NGP areas.

Average Expenditure

Regarding the length of treatment for children suffering from diarrhoea in NGP, out of 29 cases, 28 children have undergone treatment for one day, while in non-NGP areas, out of 25 children, 22 have undergone treatment for one day. The average expenditure incurred for treating all diarrhoeal patients from non-NGP villages amounts to ₹886 and to ₹679.3 for NGP areas.

In NGP areas, out of 5 pneumonia patients, 2 of them have undergone treatment for 3 days, whereas, in non-NGP areas, out of 8 pneumonia cases, 4 of them have been reported undergoing treatment for 1 day. The average expenditure for the treatment of pneumonia cases in non-NGP comes to ₹6,550 and to ₹8,800 in respect of NGP areas.

In NGP areas, out of 2 malaria cases, one malaria affected child is reported to have undergone treatment for 4 days, while in non-NGP areas, out of 5 children, 3 of them are reported to have received treatment for 3 days. The average expenditure incurred on malaria treatment in NGP villages amounts to ₹5,500 and in the case of non-NGP areas to ₹4,460.

In NGP areas, out of 30 cases of cold, cough and fever, 28 are reported to have received treatment for one day, while in non-NGP areas, out of 33 cases, 26 are reported to have received treatment for one day. The average expenditure incurred for treatment of cough, cold and fever together cases in NGP areas amount to ₹876.7 and in the case of non-NGP areas to ₹1,318.

In non-NGP areas, only one jaundice patients is reported to have undergone treatment for 4 days, with the treatment cost amounting to ₹10,000. There is only one case of skin disease reported and the expenditure on treatment works out to ₹2,750. The severity of illnesses is found more in non-NGP areas (Tables 3.39 and 3.40).

Table 3.39

Number of Children Reported Suffering from Illnesses in NGP Villages: MP

	Type of Disease	Affected	Average Episodes	Treatment Place				Length of Treatment (days)				Average Expenditure (₹)
				SC	PHC	DH	Private	1	2	3	4	
Water borne diseases	Diarrhoea	29	4.14	1	3	0	25	28	0	0	1	679.3
	Dysentery	2	5	0	0	0	2	2	0	0	0	5250
Worm infections	Amebiasis	2	2.5	0	1	0	1	1	0	0	1	7500
Mosquito bite infections	Fever	26	3.77	0	0	0	26	23	1	2	0	1819
	Malaria	2	1.5	0	0	0	2	0	1	0	1	5500
Others	Cold and fever	15	2.93	2	2	0	12	14	1	0	0	1413
	Cough and cold	27	3.56	0	1	1	25	20	6	0	1	888.9
	Cough, cold, fever	30	4	0	2	0	28	28	0	0	2	876.7
	Pneumonia	5	2.6	0	0	0	5	1	1	2	1	8800
	Others											
	Grand total	139	3.67	2	9	1	127	118	10	4	7	1619

A Profile of Household Diseases

The field based information reveals various diseases prevailing in NGP and non-NGP villages. The respondents are prone to multiple diseases mostly on account of their consuming unsafe water. The diseases are mostly non-chronic in nature and can be arrested through supplying of nutritional food, safe drinking water and sanitation and a conducive environment. The diseases like cold and cough, viral fever, body pain, vomiting, gastric enteritis, malaria, typhoid, loose motion, diarrhoea, jaundice, indigestion, can be overcome by enabling people to have access to safe drinking water and sanitation and also organizing health campaigns on a regular basis. The prevalence of chronic diseases like cancer, sugar, blood pressure, cardiac/heart, knee pain, spinal cord problems etc. is mostly

Table 3.40

Number of Children Reported Suffering from Illnesses in Non-NGP Areas: MP

	Type of Diseases	*Affected*	*Average Episodes*	*Type of Treatment Facility*					*Length of Treatment (Days)*				*Average Expenditure ₹*
				SC	*PHC*	*DH*	*Pri-vate*	*PHC & Private*	*1*	*2*	*3*	*4*	
Waterborne diseases	Diarrhoea	25	2.8	0	5	1	18	1	22	1	1	1	886
	Dysentery	1	4	0	1	0	0	0	1	0	0	0	300
	Jaundice	1	2	0	0	0	1	0	0	0	0	1	10000
	Amoebas	1	2	0	0	0	1	0	1	0	0	0	400
Worm infections	0	0	0	0	0	0	0	0	0	0	0	0	0
Mosquito bite infections	Fever	37	3.16	0	8	0	28	1	35	1	0	1	710.8
	Malaria	5	3	1	0	0	4	0	2	0	3	0	4460
Others	Cold and fever	5	3.2	0	0	0	4	1	4	1	0	0	1400
	Cough and cold	10	3.3	0	0	0	9	1	8	2	0	0	870
	Cough, cold, fever	33	3.61	0	2	0	29	2	26	5	1	1	1318
	Pneumonia	8	1.75	0	0	0	8	0	4	0	2	2	6550
	Skin disease	2	4	0	0	0	2	0	1	0	0	1	2750
	Others	3											
	Total	131	3.1	1	16	2	106	6	105	10	8	8	1604

Table 3.41

Illness Episodes Encountered, Treatment Sought and Expenditures Incurred by Households in NGP Areas: MP

	Type of Diseases	Affected	Average Episodes	Type of Treatment Facility						Length of Treatment (Days)				Average Expenditure ₹
				SC	PHC	CHC	DH	Private	PHC & Private	Less than 5	5-10	11-15	Above 15	
Waterborne diseases	Typhoid	7	1	0	1	0	0	6	0	1	1	0	5	11285.7
	Diarrhoea	6	1	1	0	0	0	5	0	2	3	1	0	4883.33
	Jaundice	1	1	0	0	0	0	1	0	0	0	0	1	3000
	Cold and cough	11	1.45	0	0	0	0	11	0	7	3	0	1	2618.18
	Digestion problem	2	1.5	0	0	0	0	2	0	1	1	0	0	12500
	Vomiting	2	1.5	0	0	0	0	2	0	0	1	0	1	12500
Mosquito bite related diseases	Fever	40	1.25	1	3	1	0	34	1	27	12	0	1	5595
	Body pains	2	1	0	0	0	0	2	0	2	0	0	0	2900
	Malaria	6	1.17	0	1	0	1	4	0	2	1	0	3	3750
	Pneumonia	4	1	0	0	0	0	4	0	1	0	0	3	75500
	Others	25												
	Fits	1	1	0	0	0	0	1	0	1	0	0	0	1000
	Stomach pain	19	1.32	0	0	0	0	18	1	9	8	1	1	3200
	Grand total	126	1.25	2	6	1	1	114	2	59	40	3	24	9616.67

Table 3.42

Illness Episodes Encountered, Treatment Sought and Expenditures Incurred by Households in Non-NGP Areas: MP

	Type of Diseases	*Af-fected*	*Average Epi-sodes*	*Type of treatment facility*						*Length of Treatment (Days)*				*Average Expendi-ture ₹*
				SC	*PHC*	*CHC*	*DH*	*Pri-vate*	*PHC & Private*	*>5*	*6-10*	*11-15*	*Above 15*	
Waterborne diseases	Typhoid	1	4	0	0	0	0	1	0	0	0	0	1	5000.00
	Jaundice	1	1	0	0	0	0	1	0	1	0	0	0	8000.00
	Fever	6	7.17	0	1	0	0	4	1	4	0	0	2	1416.67
	Motions	1	2	0	0	0	0	1	0	0	0	0	1	25000.00
	Cold and cough	5	3.2	0	0	0	0	4	1	4	1	0	0	460.00
	Diarrhoea	6	2	0	1	0	1	4	0	3	2	0	1	1383.33
Worm infection diseases	Amebiasis	0	0	0	0	0	0	0	0	0	0	0	0	
Mosquito bite related diseases	Fever	39	3.21	1	5	0	0	30	3	24	13	0	2	1846.15
	Fits	1	3	0	1	0	0	0	0	0	1	0	0	1000.00
	Malaria	6	1	1	1	0	0	4	0	1	4	0	1	2833.33
	Knee pain and stomach pain	15	4	0	1	0	0	13	1	11	3	0	1	2633.33
	Pneumonia	3	1.67	0	0	0	0	3	0	1	1	0	1	5833.33
	Others	18												
	Total	104	3.35	3	11	1	1	82	6	55	30	1	18	3419.23

common in both the areas and people also face problems due to lack of proper medical facilities in the hinterland and medical assistance at the appropriate time.

In this section, we have discussed illness episodes encountered, treatment sought and expenditure incurred by households in NGP and non-NGP villages as follows.

Illness episodes: A total of 126 persons, out of a sample of 765 in NGP villages and 104 persons out of a sample of 653 in non-NGP villages have reported facing illnesses in non-NGP areas, only one person is reported to have suffered from typhoid, while 7 persons in NGP areas. Six diarrhoea cases in each of these areas have also been reported. Many persons have reported suffering from cough and cold in both the areas (5 in non-NGP and 11 in NGP). Similarly, many persons are reported to have suffered from fever in both areas (39 non-NGP and 40 NGP).

Treatment source: Majority of those who had fallen ill are reported to have sought treatment from private hospitals with a very few approaching government health facilities.

Length of treatment and expenditure: In both NGP and non-NGP areas, the length of treatment for typhoid has been reported exceeding 15 days for 6 persons. The average expenditure incurred in the case of typhoid patients in NGP as reported by respondents is as high as ₹11,285.71, whereas in non-NGP areas, it is ₹5,000.

In NGP and non-NGP areas, for cold and cough, less than 50 per cent are reported to have undergone treatment for more than 6 days. The average expenditure incurred for cold and cough treatment in NGP areas amounts ₹2,618, whereas in non-NGP areas, it is ₹460.

The length of treatment for diarrhoea as reported by the respondents in NGP, in case of 3 patients the length of treatment for diarrhoes is 5-10 days while in non-NGP it is less than 5 days for 3 patients. The average expenditure incurred on diarrhoea patients in NGP areas amount to ₹4,883.33, whereas in non-NGP area to ₹1,383.33.

In non-NGP areas, the average number of episodes for fever is high (10.38 in non-NGP and 1.25 in NGP) and comparatively more number (12 in NGP and 13 in non-NGP villages) of them are reported to have undergoing treatment for more than 6 days. The average expenditure incurred on fever in NGP areas amount to ₹5,595, whereas in non-NGP areas, it is ₹3,262.82.

Thus, there is not much of a difference between these two areas in terms of persons affected by different diseases (Tables 3.41 and 3.42).

Summary of Focus Group Discussions (FGDs) Held in Both NGP and Non-NGP Villages in MP

Main Source of Water

- In Pitgara, Nagora, and Raipura NGP villages, the main source of water is hand pumps whereas in Morgadhi, Pedwi and Nawadpura villages the main source of water is taps.
- In Bilda, Amalijhuma and Hanumantya non-NGP villages, the main source of water is hand pumps, whereas in Sala, Rampura and Kachibaroda villages, it is taps.

Adequacy of Water Supply

- There is an adequate water supply in Pitgara, Nagora, and Raipura (NGP) villages. Whenever public hand pumps go dry, people make use of neighbours hand pump for water. During summer season, people purchase water for both domestic and drinking purposes. There is no piped water supply from the *panchayats* concerned. During rainy season, water becomes yellowish and muddy and, therefore, will not be safe for drinking. Morgadhi village receives an adequate water supply through Narmada river line. Big farmers have their own personal tap connections and pay water charges at the rate of ₹60 per month. In Nawadpura and Pedwi villages, people get a sufficient amount of water throughout the year.
- In Hanumantya, Bilda and Amalijhuma (non-NGP) villages, every one has a ready access to water. However, they get water from unhygienic and unprotected wells and hand pumps. But, during the summer season, there is a problem of water supply experienced in Bilda, Amalijhuma and Hanumantya villages (NGP) as the hand pumps go dry, while in Sala, Rampura and Kachhibaroda villages, there is an

adequate water supply during all the seasons. However, in summer, if there is any problem of water supply for one or two days, they get water from hand pumps.

Quality of Water

- In Pitgara, Nagora and Raipura NGP villages, the quality of drinking water is good though hard in nature. Some families (3-4 families in the villages) have water purifier facilities at home. In Nawadpura village, water quality is good in terms of taste and color as the GP has well sprayed with bleaching powder and chlorine.
- In Sala, Rampura and Kachhibaroda villages, water quality is good. But during the rainy season, water becomes yellowish in colour and hence, people strain water using a cloth.

Water Costs

- Whenever there is shortage of water during the summer season in Pitgara, Nagora and Raipura NGP villages, villagers pay ₹200 to the GP for arranging water from outside. In Pedwiand Morgadhi villages, villagers pay ₹50 per person (including children) to the GP for getting water. Local authorities are not properly addressing this water issue. In Nawadpura village, till now, villagers have not paid any tariff for water supply. But recently, the villagers discussed and debated the water problem in the Gram sabha and agreed to pay a nominal amount to GP.
- In Bilda, Amalijhuma and Hanumantya non-NGP villages, people don't pay for water as there is no water supply from the panchayats. Only in Rampura village, each household pays water charges (₹70 per month) to the panchayat. In Sala village also, the *panchayat* collects some amount from each family.

Health Problems Due to Contaminated Water

- In Pitgara, Nagora and Raipura-NGP villages, some respondents have reported that they are prone to suffering from fever and vomiting, especially children (0-5 years). They take treatment from ASHA and, sometimes, they go to nearby hospitals for treatment. Health problems like cold, amebiasis and dysentery are faced by the children during rainy season. Due to unsafe water, people tend to suffer from diseases like fever, dysentery, stomach ache, vomiting etc. In Pedwi and Morgadhi villages, people tend to suffer from stomach ache due to unsafe drinking water during the rainy season, while children suffer from cold, cough, fever and stomach ache. In Nawadpura village, people get affected by diseases like fever, cold, coughs etc. They go to Nisarpur PHC and, sometimes, to Badwani on doctor's reference. PHC is very far from their village. ANM visits village and conducts meetings. During these visits, she advises people on various health issues.

- In Bilda, Amalijhuma and Hanumantya—non-NGP villages, the main health problems are fever, cold and stomach ache. There are four Bengali doctors (quacks) in Bilda village and some people prefer to approach them. In Amalijhuma village, people go to doctors at Badwani, or Doganura village for treatment. They prefer private hospitals for treatment because they think that in government hospitals, one has to run around for getting treatment. In Hanumantya village, ANM or ASHA does not visit the village/locality regularly. In Sala, Rampura and Kachhi Baroda villages, people are prone to suffering from fever, jaundice and stomach ache and sometimes, children suffer from fever due to consumption of muddy water. For seeking treatment, people go to the doctor in Dharampuri or Sundrel, which is very far from their residence.

Sanitation

- In Pitgara (Harijanamohallah), Nagora, Raipura and Morgadhi-NGP-villages, some households do not have in-house toilets. Even those who have toilets do not use them, due to their bad condition. Many households in Nawadpura and Pedwi villages have access to in-house toilets.

- In Bilda, Rampura, Amalijhumma and Hanumantya—non-NGP villages, people do not have in-house toilets facility. But, some rich households enjoy in-house toilet facility with adequate water supply. However, most of the households prefer open defecation.

Open Defecation

- A maximum number of people in NGP villages practice open defecation as they don't like the toilet infrastructure, and some people take recourse to open defecation whenever pits gets filled up.
- A majority of the households in non-NGP villages go for open defecation.

Problems Due to Open Defecation

- People practicing on open defecation in NGP villages tend to face problems during night as they have to go to far off places. Moreover, there are wild animals in the villages. In some cases they feel shy. There is a problem of mosquitoes and flies over there. These problems are more during rainy season.
- In Amalijhumma—non-NGP village, people are used to open defecation. They do not agree that they might face health problems due to open defecation. In Hanumantya and Kachhibaroda villages, during rainy season, they have to search for safe places for open defecation. Children are more vulnerable to health problems like dysentery, fever, cold and cough, and suffer for 2-3 days.

Working of Committees

- In both NGP and non-NGP villages, villagers have no idea about the presence of VWSC.
- In Pitgara, Nagora, Rajpura and Pedwi NGP villages, people are not aware of VWSC. ASHA supply chlorine pills and

bleaching powder for cleaning water. ASHAs conduct meetings at Arogya kendras and counsel people on health issues during home visits. There is no ASHA worker in Nawadpura village.

- In Bilda, Amalijhuma, Sala and Kachhibarod—non-NGP villages, people are not aware of these committees. In Rampura village, there is no ASHA worker. Only ANM and MPW (multipurpose worker) attend to the health issues of the villagers during immunization day; each ANM is responsible for three villages (including their village). Sometimes, they escape from work on the pretext that they have attended to work in other villages. In Hanumantya village, one ASHA worker has recently joined but many people do not know about her presence.

Suggestions to Improve the Functioning of VWSCs and VHSCs in the Villages (NGP and Non-NGP)

- In NGP villages, participants in FGDs feel that there is a need for improving the functioning of VWSCs and VHSCs. In Pitgara village, participants expect the committees to ensure toilet facility to every house in the colony. In Nagora village, they have suggested that the SC be restored which now remains abandoned so that the doctor reside in the village for attending critical health problems. In Raipur village, the distance to SC is 2 km (this SC is situated in another colony of the village). In Morgadhi village, some women who attended the FGDs have complained that the GP is not taking up the cleaning of streets and that drainage is in bad condition and also they have to clean the streets by themselves. FGD participants expect a sweeper tobe appointed for cleaning and the laying cement roads in the village by the GP. The existing toilets should be repaired and new toilets should be constructed. In Pedwi village, there should be a clinic for children. The GP should take up cleaning of streets in the village to keep the surrounding clean on daily basis. They should provide hospital facility in the village. Every household should have tap connection. The GP should provide new wells. In Nawadpura village, toilets should be constructed in every household.

- In Bilda non-NGP village, GP should take up the cleaning of drainage system and streets. SCs should function regularly; toilet facility for every house should be provided as wild animals roam around in the village. In Kachhibaroda village, a worker should be appointed at the village level for cleaning streets and the drainage system should be maintained properly. In Sala village, toilets constructed by the government are in good condition. It would be better if the panchayat undertakes the responsibility of cleaning toilets by appointing workers. In Rampura village, the GP should construct roads and drainage system in the village besides providing tap connection for all the houses. Surroundings of hand pumps should be kept clean. The SC should be kept opened throughout the day. A doctor should be available always in the village. In Amalijhuma village, the water tank should be cleaned regularly to ensure clean and safe water to the people. The availability of transportation facilities to reach the nearby hospital needs to be ensured. In Hanumantya village, the GP should lay cement roads in each ward, construct a good drainage system, provide public toilets and individual toilets in the houses and establish SCs in the village.

Summary

As discussed in the introduction chapter, positive health outcomes depend on different interventions: direct health interventions and other institutional interventions. An effective implementation of the health programmes can lead to the low morbidity levels, reduced episodes of diarrhoea and a better nutrition absorption among children etc., in the villages. However, positive health outcomes largely depend on how the interactions take place among various institutions, processes and different programmes. It is the social determinants considered in this study, such as the availability of quantity and quality water, and sanitation facilities (in this case lavatories and their proper use by households), allied household level hygiene practices conducive to promoting health, personal hygiene that ultimately shape the health outcomes. Providing such an enabling environment is possible indirectly through a catalytic process in terms of close interactions among various institutions, processes and different programmes at the local level. The study is basically explanatory and analytical in nature.

Here an attempt is made to analyse and understand the health outcomes of individuals and households as these are linked to their access to water and sanitation facilities as well as health and hygiene awareness, and practices at the individual and household levels.

Water: In MP, 6 NGP villages and 6 non-NGP villages were selected for the study. From these villages, altogether, data was collected from 240 households. A comparison of NGP villages and non-NGP villages indicates that the pure water availability is more in NGP villages. The status of volume of water availability to the respondents amounts to the extent of 76.2 per cent for daily needs. The availability of water (both potable and non-potable) is a concern in both NGP and non-NGP areas. However, 61.5 per cent of the respondents have been experiencing inadequacy of water supply during the past one year. As per more than half of the respondents in respect of both the NGP and non-NGP villages, water is pure, but some respondents are not satisfied with the quality of water. Although some people clean drinking water vessels daily, many clean once in 2 days. Hygiene related practices relating to drinking water are somewhat better in NGP areas. For example, the respondents who use special tumblers attached with long handles (to draw water from drinking water containers) to avoid contamination through hand touching is more in NGP areas. Similarly, a large number of households in NGP areas treat drinking water before drinking as a safety measure.

Drinking Water Status in Schools: Regarding drinking water supply in government schools, out of 12 schools 6 use hand pumps (50%) and bore well (25%) as the main source in both NGP and non-NGP villages. Only one school each in non-NGP and NGP have tap connection for supplying of safe drinking water.

Regarding giving clean and safe drinking water to their children, according to a majority of the respondents in both NGP (89.9%) and non-NGP (82.2%) villages, they strain water, using a cloth. There is only a 7 per cent difference between the NGP and non-NGP villages in this respect, while only 9 of the respondents in both NGP and non-NGP villages boil water before giving to children.

An overwhelming number of the respondents in both NGP (99.2%) and non-NGP (98.3%) villages cover cooked food with a lid as away of maintaining hygiene and preventing occurrence of diseases.

More than half of the respondents help children take bath every day, especially in NGP villages, while in both NGP and non-NGP villages, children wash their hands before taking food.

Housing: Most of the respondents in NGP and non-NGP areas live in huts or kutcha houses. The general perception is that people living in semi-*pucca* or *pucca* houses are relatively better off and can afford better sanitation facilities.

Sanitation: Perhaps due to the typology of housing, only 140 households in both NGP and non-NGP areas (96 from NGP and 44 from non-NGP) have access to toilets.

A majority of the households have in-house toilet facility for their children use in NGP villages, whereas in Non-NGP villages, children practice open defecation. A majority of the respondents are not aware of diseases children may be vulnerable to, with open defecation habits and non-use of toilets by children. The availability of water for sanitation purpose is a problem in both NGP (12.6%) and non-NGP (17.4%) villages. Hence, only 38 per cent of the respondents in both NGP and non-NGP villages clean toilets daily, while about 12 per cent of the respondents clean their toilets once in a month, because of insufficient water supply which is unsafe for cleaning toilets. Those who are using toilets (96 in NGP and 44 in non-NGP) face a number of problems like flies and mosquitoes, foul smell, super structure does not ensuring privacy, over flowing of pits etc. Therefore, a mere accessibility to toilet facility does not ensure better health outcomes. A great majority of respondents are affirmative regarding, the availability of hand wash near toilets in both NGP and non-NGP villages,

In non-NGP villages, all 6 schools use pit latrines, while in NGP villages, 4 schools use pit latrines. Piped sewer system and septic type of latrines are used by NGP schools. However, in NGP villages, there are separate toilets for girls, boys and teachers. The school management engages/hires the services of private individuals for cleaning toilets, but services are not up to the mark, as the management keeps failing to pay the charges regularly for the services rendered by individuals.

Here an attempt is made to understand the role of institutions like VWSC, VHSC, CBOs etc. and the processes that were adopted to achieve ODF villages and provision of clean drinking water to all. An attempt is also made to understand the role of the institutions and the processes that hindered in achieving the ODF status of the villages and provision of clean drinking water to all.

The role of institutions: There are some local level institutions for facilitating a better implementation of water and sanitation programmes at the grass roots level such as VWSC and VHSC. However, most of the committees do not meet regularly, especially in non-NGP villages. Many times, minutes of the meetings are not recorded and the issues related to water supply are discussed orally, especially in non-NGP villages. Similarly, many of the villagers are not aware of the existence of VHSC/constitution of VHSC and their functioning at the village level.

Illnesses among children: In MP, a relatively more number (139) of children suffer from various ailments in NGP areas as compared to non-NGP areas (131), although the difference is only marginal. In both the areas, a majority of the suffering children are taken to private medical practitioners/private hospitals for treatment. In non-NGP areas, 25 diarrheal cases have been reported as against 29 in NGP areas. In both the areas, all the child patients are reported to have got treatment from private hospitals. In non-NGP areas, 5 malaria cases and in NGP areas, only 2 malaria cases have been reported. In non-NGP areas, 33 cases of children suffering from cough, cold and fever in comparison to 30 in NGP areas have been reported. In NGP areas, only 2 cases of dysentery, while only one case of dysentery in non-NGP areas have been reported.

Illnesses among adults: Regarding health problems of adults in MP, 126 cases in NGP areas and 104 cases in non-NGP areas have been reported. In both the areas, most of the patients are reported to have received treatment from private clinics. In non-NGP areas, 50 cases of adults suffering from cough, cold and fever (multiple health problems) as compared to 51 in NGP areas have been reported with a majority of them receiving treatment from private clinics. However, no dysentery cases have been reported from both NGP and non-NGP areas.

On the whole, it appears that, though in some respects things are better off in NGP villages, the conditions in NGP villages are not so different from non-NGP villages. As a result, there is not much of an impact of these water and sanitation programmes on better health outcomes in terms of household level episodes of morbidity in the study villages of MP.

4

Water and Sanitation Programmes and their Effects on the Health Status of Communities in Odisha

Introduction

The population of Odisha, according to the 2011 Census, stands at about 41 million, making it the 11th most populous state in India. The state makes up about 3.4 per cent of the country's population, as against about 3 per cent during the last Census in 2001. The state is spread over an area of about 150,000 sq. km, making it the 9th largest state in the country in terms of area. The density of population per sq km is about 260, fairly below the national average.

The state has a growth rate of about 14 per cent which is below the national average by a good three points. The overall literacy rate in the state is about 73 per cent, a tremendous improvement in the last few years due to consistent efforts on the part of the government. The sex ratio in Odisha is a healthy 978 per thousand males and the state has never witnessed a major fall in this figure. Bhubhaneshwar happens to be the capital city, and is also the largest city in the state of Odisha. In total, Odisha comprises 30 districts.

A Basic Demographic Profile of Odisha

As per details from Census 2011, Odisha has population of 4.2 crores, an increase from figure of 3.68 crore in 2001 Census. Total population of Odisha as per 2011 Census is 41,974,218 of which male and female are 21,212,136 and 20,762,082 respectively. In 2001, total population was 36,804,660 in which males were 18,660,570 while females were 18,144,090 (Census 2011).

Total population growth in this decade was 14.05 per cent while in 2001 Census, it was 15.94 per cent. The population of Odisha forms 3.47 per cent of India in 2011. In 2001, the figure was 3.58 per cent.

Literacy Rate 2011

The overall literacy rate in Odisha has seen an upward trend at 72.87 per cent, as per 2011 Population Census. Of that, male literacy rate stands at 81.59 per cent, while female literacy rate at 62.46 per cent. Comparatively, in 2001, the overall literacy rate in Odisha was 63.08 per cent with males and females accounted for 71.28 per cent and 50.51 per cent respectively. In actual numbers, total literates in Odisha stand at 26,742,595 of which males are 15,089,681 and females 11,652,914 (see Table 4.1).

Sex Ratio

The sex ratio in Odisha is 978 i.e. for every 1000 males, much above the national average of 940, as per census 2011. In 2001, the sex ratio was slightly less at 972 per 1000 males in Odisha (Census 2011).

Situation in the Study Districts and Villages

In Remuna block, the share of households with tap water from treated source amounts to 7.3 per cent; 1 per cent of the households are connected with tap water from untreated source; 57.3 per cent are dependent upon hand pumps, followed by 27.6 per cent of the households that are dependent upon tube wells/bore wells, only 2.4 per cent are dependent on rivers/canals and 1.7 per cent are dependent on uncovered wells (Census 2011).

In Nilagiri block, 4.3 per cent of the households are connected with tap water from treated source; 4.8 per cent of the households with untreated tap water source; a high per cent of households (55.6%) are dependent on hand pumps, followed by 18.3 per cent on tube wells/bore holes and 10.7 per cent of households upon uncovered wells; 1 per cent households on covered wells; 1.3 per cent of households on rivers/canals and a less than one per cent of households are dependent on springs (Census 2011).

In Baliapal block, 8.2 per cent of the households are connected with tap water from treated water source; 1.8 per cent of the household are connected with tap water from untreated water source; a large number of households (79%) are dependent on hand pumps; 8 per cent of the households on tube wells/bore holes and only 0.4 per cent are dependent on covered well (Census 2011).

In Patana block, 1.7 per cent of the households are connected with tap water from treated source, 3.6 per cent with tap water from untreated source; a high per cent of households (60.0%) are dependent upon hand pumps, 6.5 per cent on tube wells/bore holes and 2.1 per cent, 1.4 per cent, 1.5 per cent are dependent on springs, rivers/canals, tanks/ponds/lakes respectively (Census 2011).

As per Census 2011, 6 per cent of the households are connected to tap water from treated source in Basta Mandal (rural areas) while 2.3 per cent of the households to untreated source; 69 per cent of the households are dependent on hand pumps, 19 per cent on tube wells/bore wells (Census 2011).

Table 4.1

Demographic Profile of Odisha

States	*Odisha*					
	2001			*2011*		
	Male	*Female*	*Total*	*Male*	*Female*	*Total*
Population	18,660,570	18,144,090	36,804,660	21,212,136	20,762,082	41,974,218
Literacy rate	71.28%	50.51%		81.57%	62.48%	

Source: Census 2011.

In Jaleswar block, the households that are connected to tap water from treated source come to 9 per cent, while 0.7 per cent of the households to untreated tap water; nearly three fourth i.e.73 per cent of the households depend on hand pumps; 13 per cent on tube wells/bore holes and one per cent depend on rivers/canals (Census 2011).

Water and Sanitation

Sources of Drinking Water

As per Census 2011, the share of households connected with tap water from treated source in Odisha rural areas amounts to 3.9 per cent, while 3.5 per cent households are connected with untreated sources about (18%) of the households depend on uncovered wells, 44.68 per cent with hand pumps; about 20.2 per cent depend on tube wells/bore wells (Census 2011).

Sanitation

In rural areas of Odisha, only 14.1 per cent per cent of the households have access to in-house latrine facility, while the remaining 85.9 per cent of the households don't have latrine facility. The type of latrine facilities that the households have within the premises are: 55.1 per cent have septic tanks, 6.2 per cent piped sewer system under pit latrine, 13.7 per cent have slab/ventilated improved pits, 10.2 per cent have no slab/open pits. Among the households without latrine facility, 98.6 per cent practice open defecation and the remaining 1.4 per cent depend on public latrines (Census 2011).

In Baliapal block (in rural areas), 68.5 per cent of the households have no toilet facility within the household premises, while only 31.5 per cent of the households have access to in-house toilet facility. The details about the type of latrine facility within the premises among the households are: 5.6 per cent of the households have access to septic tanks, while 0.8 per cent of the households have piped sewer system and households having flush/pour-flush latrines account to 5.3 per cent. Under pit latrine, 10.8 per cent of the households have access to with-slab/ventilated improved pit latrines, while households having without-slab pit latrine constitute 8.6 per cent and 68.2 per cent of the households practice open defecation and the remaining 0.4 per cent of the households are dependent on public toilets (Census 2011).

Figure 4.1

Main Source of Water in Odisha: Overhead Tanks, Directly Connected to Pipelines

Source: Field Work.

In Remuna block (in rural areas), 80.3 per cent of the households do not have access to in-house toilet facility, while only 19.7 per cent of the households have in-house toilet facility. The details about the type of latrine facility available within the premises are households having septic tanks constitute 7.1 per cent, those with piped sewer system constitute 1.7 per cent and those with flush/pour flush latrines 3 per cent. Under pit latrines, 4.6 per cent of the households have with-slab ventilated improved pits, while households without-slab pit latrines make up 2.1 per cent and 75.4 per cent of the households practices open defecation and the remaining 4.9 per cent of the households are dependent on public toilets (Census 2011).

In Niligari block (in rural areas), 81.7 per cent of the households do not have in-house toilet facility and only 18.3 per cent of the households have access to in-house toilets. The details about the type of latrine facility, available within the household premises are: septic tank having households amount to 6.3 per cent; 0.7 per cent of the households have piped sewer system and those flush/pour flush latrine are 1.7 per cent. Under pit latrine, 6.1 per cent of the households have with-slab ventilated improved pits; those with without-slab pit latrines make up 2.3 per cent and 80.8 per cent of the households practice open defecation and the remaining 1.7 per cent of the households are dependent on public toilets (Census 2011).

In Patana block (in rural areas), 90.5 per cent of the households do not have in-house toilet facility and only 9.5 per cent of the households have access to in-house toilet facility. The details about the type of latrine facility available within the households premises are: households having septic tank make up 6.6 per cent; 0.1 per cent of the households have piped sewer system and those with flush/pour-flush latrines constitute 0.5 per cent. Under pit latrine, 1.2 per cent of the households have with-slab ventilated improved pits; 90.4 per cent of the households practices open defecation and the remaining 0.1 per cent are dependent on public toilets (Census 2011).

In Basta block (rural areas), 26.7 per cent of the households have in-house toilets and 73.3 per cent of the households go without latrine facility. The details about the type of latrine facility available within the household premises are: 21.6 per cent have latrines with septic tanks; 4.4 per cent have piped sewer system and another 12 per cent have other systems under flush/pour-flush latrines. Under pit latrine, 32.6 per cent of the households have access to with-slab/ventilated improved pits, while, 27 per cent have without-slab open pits. Among the households that don't have latrine facility, 97 per cent practice open defecation and the remaining 3 per cent depend on public latrines (Census 2011).

In Jaleswar block (rural areas), 30 per cent of the households have in-house latrine facility and 70 per cent of the households go without latrine facility. The details about the type of latrine facility available within the household premises are: 23.4 per cent have latrines with septic tanks; 4.2 per cent have piped sewer system and another 20 per cent have other systems under flush/pour-flush latrines. Under pit latrine, 18.4 per cent of the households have with-slab/ventilated improved pits; 31.6 per cent have without-slab open pits. Among the households without latrine facility, 96 per cent practices open defecation and the remaining 4 per cent are dependent on public latrines (Census 2011).

Table 4.2

Situation of Drinking Water and Water and Sanitation in Odisha

	Water		
Drinking Water	*Urban*	*Rural*	*Total*
Within the premises	862,426	1,299,251	3,418,321
Near the premises	374,820	3,706,267	2,161,677
Away	279,827	3,138,494	4,081,087
Total	1,517,073	8,144,012	9,661,085

Source: Census, 2011.

	Sanitation		
Sanitation Facility	*Urban*	*Rural*	*Total*
Having facility	982,744	146,552	2,129,296
Not having facility	534,329	6,997,640	7,531,789
Total	1,517,073	8,144,012	9,661,085

Source: Census, 2011.

A Profile of the Study Villages in the District

To understand the public policies and governance structures, it is necessary to study the environmental factors like location, socio-economic structure, infrastructural facilities, civil society bodies, etc, for, they wield a considerable influence on the performance of the development programme. The social development policies such as provision of protected drinking water and sanitation have to operate in the context of geographical, socio-cultural and political configuration of the society or the areas where they are being operated. The interaction between the environment and administration devised for the implementation of policies is very important, especially at the grass-roots levels. In this context, an attempt is made to examine some of the social determinants and infrastructural facilities available in the villages which were selected for the study. As mentioned earlier, the study was conducted in 12 villages (i.e. 6 Nirmal Gram Puraskar [NGP] and 6 non-NGP villages) of 2 districts i.e., Balasore and Kendujhar in Odisha state.

Location and Demographic Features of the Study Villages

The data on the availability of transportation in villages reveals that two thirds of the villages depend on private agencies/transport i.e., private buses or autos and another one fifth of them depend on the state transport corporation. However, a few of them (8.3%) do not have access to any transport facility, especially in non-NGP villages. With regard to availability of electricity 83.3 per cent of the villages have electricity connection and only 16.7 per cent of the villages go without electricity, mostly NGP villages.

Sources of Drinking Water in the Study Villages

Regarding the main source of drinking water, nearly three fourths of the villages depend on wells with overhead tanks directly connected to pipelines for procuring drinking water. Some villagers (25%) depend on open wells for this purpose, especially in non-NGP villages. Thus, the villagers in non-NGP villages are deprived of protected drinking water, which affects in turn, the health of people.

Figure 4.2

Main Source of Water in Odisha: Open Well

Source: Field Work.

Village Water and Sanitation Committees (VWSCs)

As per the National Health Mission (NHM) programme norms, in every village, there shall be a VWSC to guide and monitor water and sanitation services at the village level. In Odisha, in 6 NGP villages,

6 VHSCs (village health and sanitation committees) and 6 VWSCs have been formed, whereas in non-NGP areas, 6 VHSCs and 2 VWSCs have been formed. With regard to the awareness levels regarding the existence or constitution of VHSCs in the villages, two thirds of the villagers are aware of the presence of these committees, while the remaining one third of them are not aware of such committees. These committees have been constituted in all the NGP villages, while in non-NGP villages, only two villages have these committees as against the remaining four villages. The meetings of VHSCs are held either at the GP (*gram panchayat*) office or Indian Child and Development Services (ICDS) centres. In the case of NGP villages, meetings are held at GP office, while in respect of non-NGP villages they are held in Anganwadi centres.

The minutes of the committee meetings are recorded and preserved properly by the chairman of the committee or sarpanch of the GP and, sometimes, the employees of the Anganwadi centres or the committee members. Two thirds of committee members do not have knowledge or information about the budgetary allocation or grants by the higher authorities to VWSCs. Although the committees have been constituted in certain villages, their functioning is very discouraging as the members are not aware of their duties and powers. This is true in respects of both the NGP and non-NGP villages. Further, many villagers (87.4%) are not aware of the existence/constitution of the VWSCs especially in non-NGP villages, while in NGP villages, a majority of the people are aware of these committees.

The village *sarpanch*/secretary concerned, especially in NGP villages, have disseminated information on water and sanitation programmes through posters, hand-outs/pamphlets/wall writings/ brochures etc., while in non-NGP villages, such efforts are not visible.

Of the total villages (12), more than half of the villages i.e., 7 prepare the village health plans. From non-NGP villages, out of 6 only 2 villages prepare the plans. Further, the data reveals that in NGP villages, issues/items identified in the village plans, are brought to the notice of VHSC functionaries, whereas in three non-NGP villages, these issues are brought to the notice of the committee functionaries for resolving the problems. However, in more than three fourths of the villages (9), VHSC functionaries do not present the annual reports in the Gram Sabha of GP for discussion.

State of Water and Sanitation Facilities in Government Schools in the Study Villages

In Odisha, in all six villages, six schools and six ICDS centres are covered, whereas in non-NGP villages, five schools and five ICDS centres are covered. At the school level, observations and interviews with key authorities and teachers reveal that in NGP villages, out of six schools, three schools get drinking water from bore wells, two schools from taps and one from other sources. Out of six schools in non-NGP, three schools get drinking water from bore wells, and one each from taps, tanks and other sources.

As regards latrine facilities in schools, in NGP villages, out of six schools, 4 schools have septic tank type of toilets and two schools pit latrines, whereas in non-NGP villages, out of six schools, four schools have septic type of toilets and one each pit latrine and piped sewer system.

Out of six schools in NGP areas, three schools have separate toilets for boys and girls, whereas in non-NGP villages, in four schools out of six there is such facility. In NGP areas, only three schools have separate toilets for teachers, while all six schools do not have separate toilets for teachers in non-NGP areas.

Water Facility in ICDS Centres in the Study Villages:

As regards ICDS centres, there are six in NGP and six in non-NGP villages. In NGP villages, four ICDS centres get water from bore wells followed by one each from taps and wells. In non-NGP, out of six centres, four centres get water from 'other sources' followed by bore wells (1).

A Profile of Households and Children Behaviour Related to Water, Sanitation and Hygiene

This section is based on the data collected through household and child schedules. In this section, an attempt is made to examine the perceptions of household members representing NGP and non-NGP villages regarding the utilisation of water and sanitation programmes and their impact on the health conditions of the rural communities. Unlike the other studies, the present study adopted the programme (NGP villages) and control-group (villages without NGP) approach, covering both NGP and non-NGP village households in an attempt to juxtapose the situation for eliciting the response of people towards the programmes and their impact on the health conditions of people.

Figure 4.3

Wells within Household Premises in Odisha

Source: Field Work.

An analysis of the experiences of those who are recipients of the scheme, NGP provides insights into the operational dynamics of the programme. This also brings out the gap between promise and performance besides highlighting the strengths and weaknesses of the programmes. Similarly, an understanding and assessment of the perceptions of the non-NGP households provide a clue in identifying the causes for the non-availability of facilities offered by the government. Here, an attempt is made to analyse the socio-economic background of the respondents, representing NGP and non-NGP villages.

The distribution of the respondents reveals an overwhelming percentage of males (87.5%), while a negligible percentage of females. A similar trend, by and large, prevails in both NGP and non-NGP villages. The representation of the respondents is also the same both in NGP (50.4%) and non-NGP (49.6%) villages. The educational levels of the respondents unfold that a majority of them have completed their primary (50.8%) and secondary (30.4%) education, while a less percentage of them have pursued their graduation (12.1%) and post-graduation (1.3%). A few of the respondents (5.4%) are illiterate. However, the educational levels of the respondents representing NGP villages are relatively higher than non-NGP villages. Thus, the respondents representing NGP villages appear to have better opportunities to avail water and sanitation programmes in view of their relatively higher educational levels, as compared to non-NGP respondents (Table 4.3).

Table 4.3

Gender and Literacy Profiles of the Respondents in NGP and Non-NGP Villages: Odisha

Level of Schooling	*NGP*			*Non-NGP*			*All*		
	Male	*Female*	*Total*	*Male*	*Female*	*Total*	*Male*	*Female*	*Total*
Primary	48	7	55	49	18	67	97	25	122
Per cent	42.9	77.8	45.5	50	85.7	56.3	46.2	83.3	50.8
Secondary (6-12)	42	1	43	30	0	30	72	1	73
Per cent	37.5	11.1	35.5	30.6	0	25.2	34.3	3.3	30.4
Graduate	15	0	15	13	1	14	28	1	29
Per cent	13.4	0	12.4	13.3	4.8	11.8	13.3	3.3	12.1
Post graduate	1	0	1	2	0	2	3	0	3
Per cent	0.9	0	0.8	2	0	1.7	1.4	0.0	1.3
Illiterate	6	1	7	4	2	6	10	3	13
Per cent	5.4	11.1	5.8	4.1	9.5	5	4.8	10.0	5.4
Total	112	9	121	98	21	119	210	30	240

The analysis reveals that overall, a majority of the respondents reside in *kutcha* houses (50.4%) and huts (6.7%), and while less than one fourth and less than one fifth of the respondents reside in *pucca* and semi-*pucca* houses respectively. The data further reveals that the respondents living in *pucca* and semi-*pucca* houses belongs to NGP villages, while the respondents living in *kutcha* houses and huts come from non-NGP villages, indicating their poor economic conditions (Table 4.4).

Table 4.4

Profile of Households by Type of House in NGP and Non-NGP Villages: Odisha

Type of House	*NGP*	*Non-NGP*	*Total*
Pucca	43	19	62
Per cent	35.5	16.0	25.8
Semi-pucca	30	11	41
Per cent	24.8	9.2	17.1
Kutcha	33	88	121
Per cent	27.3	73.9	50.4
Hut	15	1	16
Per cent	12.4	.8	6.7
Total	121	119	240

Table 4.5

Distribution of the Sample Households by Type of Toilet in NGP and Non-NGP Villages: Odisha

Type of Toilet	*NGP*	*Non-NGP*	*Total*
Pit latrine	39	4	43
Per cent	32.2	3.4	17.9
Flush toilet	5	4	9
Per cent	4.1	3.4	3.8
Piped sewer system	1	0	1
Per cent	.8	.0	.4
Septic latrine	22	1	23
Per cent	18.2	.8	9.6
Open defecation	54	110	164
Per cent	44.6	92.4	68.3
Total	121	119	240

Note: Chi-value = 57.883, P= 0.

With regard to toilet facilities, the field data reveals that,overall, more than two-thirds of the respondents (68.3%) still practice open defecation and less than one-fifth (17.9%) depend on pit latrines followed by septic latrines (9.6%) in both NGP and non-NGP areas. As it is evidenced from Chi-square test, open defecation is being practiced more in non-NGP areas (92.4%) than in NGP villages (44.6%) (Table 4.5).

Table 4.6

Respondents Perceptions Related to the Use of Toilets by the Level of Education in NGP and Non-NGP Villages: Odisha

Education	*NGP*					*Non-NGP*				*All*				
	For Better Health	*Privacy*	*Persuaded by NGOs/ Govt/ Others*	*For Better Health and Privacy*	*Total*	*For Better Health*	*Privacy*	*For Better Health and Privacy*	*Total*	*For Better Health*	*Privacy*	*Persuaded by NGOs/ Govt/ Others*	*For Better Health and Privacy*	*Total*
Primary	5	14	1	3	23	0	0	1	1	5	14	1	4	24
Per cent	22	60.80	4	13	95.83	0	0	100	4.16	20.83	58.33	4	16.66	100.00
Secondary (6-12)	15	19	0	7	41	1	4	3	8	16	23	0	10	49
	37	46.34	0	17	83.67	13	50	38	16.32	32.65	46.93	0	20.40	100.00
Illiterate	0	3	0	0	3	0	0	0	0	0	3	0	0	3
	0	100.00	0	0	100.00	0.00	0.00	0.00	0.00	0.00	100.00	0	0.00	100.00
Total	20	36	1	10	67	1	4	4	9	21	40	1	14	76

Regarding the reasons for using personal toilets, more than half of them do so for having privacy and another 18.5 per cent of the respondents for maintaining good health and privacy. It is interesting to note that these responses have come mostly from the respondents with relatively higher educational qualifications. A similar trend is observed in both NGP and non-NGP villages. This also further confirms the fact that the education levels of people make them more conscious of practicing hygienic methods (Table 4.6).

Table 4.7

Details of Hand Wash Habits after Defecation in NGP and Non-NGP Villages: Odisha

Type of Hand Wash	*NGP*	*Non-NGP*	*Total*
Soap	60	39	99
Per cent	32.8	41.2	41.2
Sand/ash	46	76	122
Per cent	38.0	63.9	50.8
Plain water	15	4	19
Per cent	12.4	3.4	7.9
Total	121	119	240

Regarding washing of hands after defecation and the material used for it, more than half of the respondents use sand and another 41.2 per cent of them use soap for washing hands, while a less percentage of respondents (7.9%) use water for washing hands. There is not much of a variation between NGP and non-NGP village in this regard (Table 4.7)

Table 4.8

Water Availability and Usage (by Source) in NGP and Non-NGP Villages: Odisha

Purpose of Using Occasional Source of Water	*NGP*	*Non-NGP*	*Total*
Don't know	3	21	24
	2.5	17.6	9.6
Domestic use	105	94	199
	86.8	79.0	82.9
Toilet use	13	4	17
	10.7	3.4	7.1
Total	121	119	240

An overwhelming percentage of the respondents (82.9%) use water for domestic purpose, such as washing dishes, clothes and for bath etc, while a less percentage of the respondents (7.1%) use water for toilet purpose. This indicates that villagers (particularly in non-NGP) tend to give less priority to cleaning of toilets. This is partly due to their practice of open defecation (Table 4.8).

Table 4.9

Perceptions of the Respondents Regarding the Quality of Water in NGP and Non-NGP Villages: Odisha

Quality of Water	*NGP*	*Non-NGP*	*Total*
Pure	89	87	176
Per cent	73.6	73.1	73.3
Sometimes pure	24	32	56
Per cent	19.8	26.9	23.3
Impure	2	0	2
Per cent	1.7	.0	.8
Sometimes impure	2	0	2
Per cent	1.7	.0	.8
Mostly pure	3	0	3
Per cent	2.5	.0	1.2
Mostly impure	1	0	1
Per cent	.8	.0	.4
Total	121	119	240

Note: Chi-value =9.15, P=0.103.

With regard to the quality of water, three fourths of the respondents have mentioned that it is pure and another one fourth of them have replied that water is pure only during a certain period. The difference between NGP and non-NGP with regard to quality of water is not significant (Table 4.9). A few of the respondents (2.08%) have reported that water is impure, especially during the rainy season.

Regarding whether water is adequate to meet the daily needs, i.e., both domestic and drinking purposes, nearly half of the respondents have reported that water is insufficient, while one fourth of the respondents have replied that water availability is not adequate.

However, one fifth of the respondents have expressed that they are happy with the availability of water and able to meet their daily needs. These, respondents mostly hail from NGP villages.

Table 4.10

Details of Adequacy/Inadequacy of Water Availability for Daily Needs: Odisha

Quantity of Water	*NGP*	*Non-NGP*	*Total*
Fully sufficient	53	6	59
Per cent	43.8	5.0	24.6
Somewhat sufficient	29	39	68
Per cent	24.0	32.8	28.3
Insufficient	33	68	101
Per cent	27.3	57.1	42.1
Not at all sufficient	6	6	12
Per cent	5.0	5.0	5.0
Total	121	119	240

Thus, the analysis reveals that there is a variation in the availability of water between NGP and non-NGP villages, with the former category of villages having adequate water as compared to the latter category of villages. The policymakers have to provide an adequate attention to addressing the issue of water supply to the people living in non-NGP villages (Table 4.10).

Table 4.11

Perceptions of the Respondents Regarding the Adequacy of Water Supply During the Past One Year: Odisha

Adequacy of Water Supply	*NGP*	*Non-NGP*	*Total*
Yes	34	8	42
Per cent	28.1	6.7	17.5
No	87	111	198
Per cent	71.9	93.3	82.5
Total	121	119	240

When probed further about the availability of adequate quality water during the past one year, overall, a great majority of the respondents (82.5%) have reported that it is not adequate. This response is more from the respondents representing non-NGP villages. This indicates that the authorities concerned have neglected certain villages in terms of initiating measures to supply an adequate volume of water to the people. This is partly due to the non-functioning of GPs and the committees (VWSC and VHSC) villages in specially working for this purpose (Table 4.11).

Table 4.12

Particulars of Season-Wise Water Insufficiency (in terms of Intensity): Odisha

Seasonal Water Insufficiency	*NGP*	*Non-NGP*	*Total*
Summer	34	8	42
Per cent	100.0	100.0	100.0
Winter	0	0	0
Per cent	0.0	0.0	0.0
Total	34	8	42

Further, the respondents who have reported that the availability of water is adequate, also experience water problem during summer in both NGP and non-NGP areas (Table 4.12).

Table 4.13

Information on the Storage Capacity (Drinking Water) of Water Tanks/Drums/Vessels in the Sample Households in NGP and Non-NGP Villages: Odisha

Storage Capacity of Drinking Water	*NGP*	*Non-NGP*	*Total*
10 tubs	41	15	56
Per cent	33.9	12.6	23.3
15 pots	34	83	117
Per cent	28.1	69.7	48.8
1 drum	16	16	32
Per cent	13.2	13.4	13.3
10 buckets	30	5	35
	24.8	4.2	14.6
Total	121	119	240

With regard to the storage capacity of containers, especially for drinking water, nearly one fourths of the respondents have mentioned that they possess storage facility/ies which can preserve 10-15 tubs of water. These responses, as observed earlier also, have come more from non-NGP villages as compared to NGP villages. The respondents who use buckets for storing water mostly belong to NGP villages. This could be, as pointed out earlier, due to a regular supply of water by the GP to people (Table 4.13).

Table 4.14

Information on Practices of Drawing Drinking Water from Containers: Odisha

Method Used for taking Water from Container	*NGP*	*Non-NGP*	*Total*
Container has a tap attached	15	7	22
Per cent	12.4	5.9	9.2
With A tumbler/utensil using hand	89	74	163
Per cent	73.6	62.2	67.9
With a tumbler/utensil attached to a long handle	16	38	54
Per cent	13.2	31.9	22.5
Others	1	0	1
Per cent	.8	.0	.4
Total	121	119	240

With regard to the use of methods for taking water from containers for drinking purpose, about three fourths have mentioned that they draw water with a glass tumbler or other utensil using hand. Sometimes, this could cause the contamination of water and ill-health of household members. This situation is more visible in NGP villages. The data reveals that a little over one fifth and less than 9.2 per cent of the respondents use glass tumblers attached with long handles and a tap attached to the container for consuming water respectively. These methods are considered as appropriate methods for consuming drinking water and for keeping good health (Table 4.14).

Almost all the respondents in both the areas have reported that they clean drinking water containers to a question on, how many times they have cleaned water containers during the past one month. An overwhelming percentage of the respondents (80%) have replied that they have been cleaning every day, while, some others have mentioned that they do so once in two days (11.2% of the respondents), and once in five days (5% of the respondents). However, an insignificant percentage of the respondents (1.2%) have replied that they clean them once in a month. As it is evidenced from Chi-square test the percentages of respondents, who clean containers every day, come more from non-NGP villages, as compared to NGP villages (Table 4.15).

Table 4.15

Distribution of Households by Frequency of Cleaning Water Containers Used for Drinking Water in NGP and Non-NGP Villages: Odisha

Cleaning Drinking Water Container	*NGP*	*Non-NGP*	*Total*
No	0	0	0
Per cent	0.0	0.0	0.0
Yes	121	119	240
Per cent	100.0	100.0	100.0
Daily	90	104	194
Per cent	74.4	87.4	80.8
Once in 2 days	19	8	27
Per cent	15.7	6.7	11.2
Once in 5 days	8	4	12
Per cent	6.6	3.4	5.0
Once in 2 weeks	2	2	4
Per cent	1.7	1.7	1.7
Once a month	2	1	3
Per cent	1.7	.8	1.2
Total	121	119	240

Note: Chi-value =7.142, P=0.129.

A less percentage of the respondents (16%) have expressed that they boil water or strain water using a cloth and that they take precautionary measures to clean water before drinking. Another one fourth of them have mentioned that, only sometimes, they go through this exercise. More than half of the respondents have

Table 4.16

Information on Treatment of Drinking Water before Consumption, by Level of Education of the Respondents: Odisha

Educational Level	*NGP*					*Non-NGP*					*All*				
	Al-ways	*Some times*	*Never*	*Not at All Done*	*Total*	*Al-ways*	*Some times*	*Never*	*Not at All Done*	*Total*	*Al-ways*	*Some times*	*Never*	*Not at All Done*	*Total*
Primary	8	17	26	4	55	14	19	32	2	67	22	36	58	6	122
Per cent	14.54	30.90	47.27	7.27	45.08	20.89	28.35	47.76	2.98	54.91	18.03	29.50	47.54	4.91	100.00
Secondary (6-12) and graduate	12	17	28	1	58	4	11	25	4	44	16	28	53	5	102
Per cent	20.68	29.31	48.27	1.72	56.86	9.09	25.00	56.81	9.09	65.68	15.68	27.45	51.96	4.90	100.00
Postgraduate	0	0	1	0	1	0	0	2	0	2	0	0	3	0	3
Per cent	0.00	0.00	100.00	0.00	33.33	0.00	0.00	100.00	0.00	66.66	0.00	0.00	100.00	0.00	100.00
Illiterate	0	3	4	0	4	0	2	4	0	6	0	5	8	0	13
Per cent	0.00	42.85	57.14	0.00	57.14	0.00	33.33	66.66	0.00	46.15	0.00	38.46	61.5%	0.00	100.00
Total	20	37	59	5	121	18	32	63	6	119	38	69	122	11	240

expressed that they do not go through this exercise. This response is relatively more from the respondents (57.9%) representing non-NGP villages and illiterates. This indicates that the educational levels of the household members play a crucial role in treating water before consumption (Table 4.16).

Further, the respondents who have mentioned that they treat water were asked to indicate which method was mostly used for treating water. Nearly three fourths of them (in both NGP and Non-NGP) have replied that they boil water, while a little over one fifth of them strain water using a cloth. Non-NGP villages were found better in treating water. A few of them (5.6%) have mentioned that they wait till sand or any other particle settle at the bottom of the pot/container. This is somewhat unsafe for drinking purpose (Table 4.16).

Table 4.17

Details of Measures Taken for Treating Water before Consumption by the Sample Households in NGP and Non-NGP Villages: Odisha

Measures	*NGP*	*Non-NGP*	*Total*
Boil water	37	39	76
Per cent	64.9	78.0	71.0
Strain water, using a cloth	14	11	25
Per cent	24.6	22.0	23.4
Let water stand and settle	6	0	6
Per cent	10.5	.0	5.6
Total	57	50	107

Note: Chi-value=5.98, P=0.05.

The respondents were asked to indicate the major problems encountered by them in accessing adequate water supply regularly in their residential/local areas. According to nearly two thirds of the respondents, the shortage of water is the main problem, a few of (6.2%) them have mentioned that they have to travel long distances for fetching water and as per some others (6.7%), adequate number of bore wells are not available in their locality. However, a little over one fifth of the respondents have expressed that they have no problems in getting water in their respective wards/localities (Table 4.18).

Table 4.18

Information on the Major Problems Related to Water Supply in the Sample in NGP and Non-NGP Villages: Odisha

Major Problem	*NGP*	*Non-NGP*	*Total*
No problem	50	4	54
Per cent	41.3	3.4	22.5
Bore wells is not available	11	5	16
Per cent	9.1	4.2	6.7
Shortage of water	53	100	153
Per cent	43.8	84.0	63.8
Long distance involved in fetching water	5	10	15
Per cent	4.1	8.4	6.2
Don't know	2	0	2
Per cent	1.7	.0	.8
Total	121	119	240

Figure 4.4

Posters on How to Make Drinking Water Safer

Source: Field Work.

Regarding major problems of drinking water faced, there are 299 multiple responses. The main problems reported are: water problem is acute during the summer season (23.74%) (Water level goes down during summer) and there are no adequate number of bore wells/

tube wells (20.4%). The main problem reported mostly in non-NGP villages is no water supply (piped water supply) in the sample villages (18.68%). Similarly, the main problem reported in NGP villages is poor electricity supply/power cut (20%). This is relevant here, as many households are connected to tap water (Table 4.19).

Table 4.19

Distribution of Respondents Reporting Major Problems Regarding Water Supply in NGP and Non-NGP Villages: Odisha

Problems	*NGP*	*Non-NGP*	*Total*
Going to another place due to non-working of bore wells	0	1	1
Per cent	0	0.5	0.3
Paying money to get water	4	1	5
Per cent	3.4	0.5	1.6
Long distance involved in fetching water	4	4	8
Per cent	3.4	2.19	2.6
During summer season, water level goes down	31	40	71
Per cent	26.4	21.9	23.74
Unprotected wells	0	2	2
Per cent	0	1.09	0.66
No public taps in the villages	0	12	12
Per cent	0	6.5	4.0
Untimely water supply	9	18	27
Per cent	7.7	9.89	9.0
Inadequate number of tube wells/bore wells	25	36	61
Per cent	21.3	19.7	20.4
No water supply (piped water supply) available in our village	1	34	35
Per cent	0.85	18.68	11.7
All parts of our village are not connected to piped water supply	9	7	16
Per cent	7.7	3.8	5.35
All households are not connected to piped water supply	6	6	12
Per cent	5.1	3.29	4.01
Poor electricity supply/power cuts	24	7	31
Per cent	20.5	3.8	10.36

Water storage facility not available in our village	0	5	5
Per cent	0	2.7	1.67
Water storage facility not available in poor families	0	1	1
Per cent	0	0.54	0.33
Contaminated cater	1	2	3
Per cent	0.85	1.09	1.0
Total	117	182	299

Table 4.20

Status of Water Supply in ICDS Centres in NGP and Non-NGP Villages: Odisha

Water Supply in ICDS Centre	*NGP*	*Non-NGP*	*Total*
Yes	63	79	142
Per cent	52.1	66.4	59.2
No	47	33	80
Per cent	38.8	27.7	33.3
Others	11	7	18
Per cent	9.1	5.9	7.5
Total	121	119	240

According to more than half of the respondents, Anganwadi centres supply water to children attending the centres, while one third of the respondents have replied negatively (Table 4.20).

The respondents were asked to reveal the advantages of having toilets within the household premises. As per some of the respondents, it is to prevent diseases and to maintain hygiene conditions, primary and social status, while more than one third of the respondents have mentioned other advantages (Table 4.21).

The respondents, using toilets were asked a question, "Were they facing any problem in using and maintaining toilets?". Many respondents (23) have replied that they encounter problems, while others have expressed that they do not have any problem with the maintenance of toilets (Table 4.22).

Table 4.21

Respondents Awareness Regarding the Advantages of Using In-house Latrine/Toilet/Lavatory Facility in NGP and Non-NGP Villages: Odisha

Advantages	*NGP*	*Non-NGP*	*Total*
Better health	23	3	26
Per cent	14.3	1.9	16.1
Privacy	13	4	17
Per cent	8.1	2.5	10.6
Ease of use	4		4
Per cent	2.5		2.5
Social status	3		3
Per cent	1.9		1.9
Better health and privacy	9	3	12
Per cent	5.6	1.9	7.5
Privacy and social status	4		4
Per cent	2.5		2.5
No toilet	23		23
Per cent	14.3		14.3
Better health and ease of use	10		10
Per cent	6.2		6.2
Others		62	62
Per cent		38.5	38.5
Total	89	72	161

Table 4.22

Respondents Encounter Problems Regarding the Use of Toilets in NGP and Non-NGP Villages: Odisha

Problem	*NGP*	*Non-NGP*	*Total*
Yes	22	1	23
Per cent	32.8	11.1	30.3
No	45	8	53
Per cent	67.2	88.9	69.7
Total	67	9	76

Table 4.23

Problems Encountered by the Sample Households Relating to Toilet Use in NGP and Non-NGP Villages: Odisha

Toilets Problems	*NGP*	*Non-NGP*	*Total*
Water not available for use/cleaning	8	1	9
Per cent	21.62	100	23.68
Flies/or mosquitoes	10	0	10
Per cent	27.02	0	26.31
Superstructure not ensuring privacy	2	0	2
Per cent	5.40	0	5.40
Foul smell	4	0	4
Per cent	10.81	0	10.81
Flooding during rainy seasons	8	0	8
Per cent	21.62	0	21.62
Difficulties for younger children to use	3	0	3
Per cent	8.10	0	8.10
Filled up pits	2	0	2
Per cent	5.40	0	5.40
Total	37	1	38

Some of the problems faced in using toilets are: mosquitoes and flies menace (26.31% of households), water not available (23.68% of households), flooding during rainy season (21% of households) and foul smell from toilets (10.52% of households). The non-availability of adequate quantity of water for cleaning toilets, an uncontrollable mosquito menace in the premise and locality, lack of privacy, bad smell due to irregular maintenance of the drainage canals, lack of connectivity of toilet pipes with drainage, blockage of toilet pipes during rainy season, unloading or irregular cleaning of toilet pits, and unsafe toilet structures are the major problems encountered by the respondents in using toilets. Some of these issues are related to the linkage of the toilets to the main drainage system in the villages and the general environment. The local government's role, in this context, is of utmost importance in controlling mosquitoes and proper maintaining the drainage system in the locality (Table 4.23).

Table 4.24

Information on the Practices of Disposing of Children Stools Post Defecation in NGP and Non-NGP Villages: Odisha

Methods of Disposal of Child's Stools	*NGP*	*Non-NGP*	*Total*
Leave it where it is	3	9	12
Per cent	1.3	3.8	5.0
Throw it on the street	80	77	157
Per cent	33.3	32.1	65.4
Throw it in the latrine	11	2	13
Per cent	4.6	0.8	5.4
Others	27	31	58
Per cent	11.3	12.9	24.16
Total	121	119	240

Note: Chi-value=49.55, P=0.

Regarding the practices of disposing of children's stools after open defecation, nearly two thirds of the respondents throw children's excreta on the street (significantly more in NGP areas) while some of them leave it where it is or throw it in the latrine. The overall situation is that the methods used by the respondents in disposing of getting children's excreta are not desirable and might cause ill-health of household members and also neighbours (Table 4.24).

Figure 4.5

Kutcha House in Odisha

Source: Field Work.

Figure 4.6

Household Toilet in a Non-NGP Village of Odisha

Source: Field Work.

Figure 4.7

Household Toilet in a NGP Village of Odisha

Source: Field Work.

Having hand wash facility near toilets indicate the importance given by the households to hygienic practices. With regard to the availability of hand wash facility near toilets, three fourths of the respondents neither have any facility to wash their hands near toilet nor do they wash hands. This situation is more visible in non-NGP villages as compared to NGP villages. We have 64 responses regarding type of hand wash post defecation (55 in NGP and 9 in non-NGP). In NGP areas, a majority of the respondents use soap (80% in NGP and 66.7% in non-NGP villages) and the remaining use plain water (Table 4.25).

Table 4.25

Information on Hand Wash Facility and Type of Hand Wash Near Toilets: Odisha

Hand Wash Facility and Type of Hand Wash	*NGP*	*Non-NGP*	*Total*
Without wash facility	66	110	176
Per cent	54.5	92.4	73.3
With wash facility	55	9	64
Per cent	45.5	7.65	26.7
Soap	44	6	50
Per cent	80.0	66.7	78.1
Sand/ash	9	3	12
Per cent	16.4	33.3	18.8
Plain water	2	0	2
Per cent	3.6	.0	3.1
Total	55	9	64

Table 4.26

Frequency of Cleaning Toilets in NGP and Non-NGP Villages: Odisha

Frequency of Cleaning Toilets	*NGP*	*Non-NGP*	*Total*
Once or more in a day	6	1	7
Per cent	9.0	11.1	9.2
At least once in a week	1	1	2
Per cent	1.	11.1	2.6
At least once a fortnight	8	1	9
Per cent	11.9	11.1	11.8
At least once a month	31	2	33
Per cent	46.3	22.2	43.4
Others	21	4	25
Per cent	31.3	44.4	32.9
Total	67	9	76

The analysis reveals that a considerable percentage (43.4%) of respondents in both NGP and non-NGP areas, have reported cleaning their toilets once in a month which makes it unsafe to use such toilets (Table 4.26).

In this section, illnesses related to unsafe drinking water and unhygienic practices followed in the study villages resulting in health problems for children such as diarrhoea, dysentery and worm infections are discussed. Further, methods followed by households for treating dehydration of children are also presented here.

Table 4.27

Treatment Seeking Behaviour of the Households with respect to Dehydration among Children in NGP and Non-NGP Villages: Odisha

Methods Followed	*NGP*	*Non-NGP*	*Total*
Provide ORS (oral rehydration solution)	58	45	103
Per cent	47.9	37.8	42.9
Provide only hot water	6	7	13
Per cent	5.0	5.9	5.4
Visit doctor/s	16	46	62
Per cent	13.2	38.7	25.8
Get medicines directly from pharmacy	0	1	1
Per cent	.0	.8	.4
Approach Village Quacks	15	0	15
Per cent	12.4	.0	6.2
Accredited Social Health Activist (ASHA)/Accredited Nurse Midwife (ANM)	8	0	8
Per cent	6.6	.0	3.3
Others	18	20	38
Per cent	14.8	16.8	15.9
Total	121	119	240

Nearly half of the respondents in NGP areas and more than one third of the respondents in non-NGP areas give ORS to children during dehydration. In non-NGP areas, nearly 38.7 per cent of the respondents prefer to visit a doctor for treatment of dehydration, while about 19 per cent of the respondents in NGP areas prefer to visit Quacks or ANMs (Table 4.27).

Table 4.28

Measures Taken by the Households to Keep Cooked Food Safe in NGP and Non-NGP Villages: Odisha

Measures	*NGP*	*Non-NGP*	*Total*
Cover cooked food	118	118	236
Per cent	97.5	99.2	98.3
Not done anything	3	0	3
Per cent	2.5	.0	1.2
Others (specify)	0	1	1
Per cent	.0	.8	.4
Total	121	119	240

The analysis of data unfolds that overall, a majority of the respondents (98.3%) both in NGP and non-NGP villages cover cooked food with a lid, while, just 1.2 per cent of the respondents do not maintain cooked food properly (Table 4.28).

Figure 4.8

ICDS Centre in Odisha

Source: Field Work.

Figure 4.9

Washing Clothes Near a Hand Pump in Odisha

Source: Field Work.

Figure 4.10

An Open Drainage Canal in a NGP Village in Odisha

Source: Field Work.

With regard to taking bath, more than half of respondents take bath everyday (66.7% in NGP and 58.8% in non-NGP villages), while the remaining one-third of them take bath once in two days. This shows that the respondents are very keen on maintaining personal hygiene (Table 4.29).

Table 4.29

Household Level Personal Hygiene Practices: Frequency of Taking Bath in NGP and Non-NGP Villages: Odisha

Frequency of Taking Bath	*NGP*	*Non-NGP*	*Total*
Once or more a day	80	70	150
Per cent	66.7	58.8	62.8
At least once in 2 days	36	41	77
Per cent	30.0	34.5	32.2
Once in a week	4	8	12
Per cent	3.3	6.7	5.0
Total	120	119	239

Awareness Level Regarding VWSC and VHSC Among the Local Population

More than one third of the respondents i.e., 61.2 per cent in NGP and 12.6 per cent in non-NGP villages are aware of VWSC. Compared to this, many respondents i.e., 98.3 per cent in non-NGP and 61.2 per cent in NGP villages are aware of VHSC (Tables 4.30).

Table 4.30

Respondents Awareness about the Existence of VWSC and VHSC in their Villages: Odisha

Existence of VWSC	*NGP*	*Non-NGP*	*Total*
Aware (VWSC)	74	15	89
Per cent	61.2	12.6	37.1
Not aware (VWSC)	47	104	151
Per cent	38.8	87.4	62.9
Aware (VHSC)	74	117	191
Per cent	61.2	98.3	79.6
Not aware (VHSC)	47	2	49
Per cent	38.8	1.7	20.4
Total	121	119	240

An Analysis of Information Based on Child Health and Hygiene

As part of the study, an attempt was also made to collect information on the various dimensions of the child health and hygiene related practices. This was done by administering an interview schedule to mothers of children aged 0-5 years. The purpose of the study/schedule was to find out health and hygiene practices followed by parents, especially mothers. The schedule covered various aspects such as practices treating drinking water, awareness about diseases/ailments children are vulnerable due to unsafe drinking water, the availability of toilet facility especially for children, diseases/ailments affecting children as a result of the practice of open defecation or non-use of toilets, personal hygiene practices, improper bathing and washing of hands and material used for washing. The child schedule also contained questions regarding the status of immunisation of children and the maintenance of immunisation card separately for each child and illness episodes undergone by children during the last one year. The data collected on these aspects helped us have an idea about not only the health status of our future generations, but also the strengths and weaknesses of the policies meant for human development, especially an improvement in the health standards of children

Regarding using drinking water for children, a majority of the respondents try to give clean water using one or the other method. In NGP areas, nearly half of the respondents let water stand and settle for some time before giving it to their children for drinking. In non-NGP areas, 42.3 per cent of them strain water, using a cloth before giving it to their children. In both the areas one-third of the respondents boil water. Overall from T. test. NGP areas found statistically significant in treating of water before drinking (Table 4.31).

Table 4.31

Distribution of Households Treating Water before Consumption and Methods of Making Water Safer for Children's Consumption: Odisha

Treatment of Water	*NGP*	*Non-NGP*	*Total*
Yes	117	104	221
Per cent	96.7	87.4	92.1
No	4	15	19
Per cent	3.3	12.6	7.9
Total	121	119	240
Practices Used for Making Water Safer for Consumption			
Boil water	45	25	70
Per cent	38.5	24.0	31.7
Strain water, using a cloth	14	44	58
Per cent	12.0	42.3	26.2
Use water filters	0	1	1
Per cent	.0	1.0	.5
Let water stand and settle	55	34	89
Per cent	47.0	32.7	40.3
Others (specify)	3	0	3
Per cent	2.6	.0	1.4
Total	117	104	221

Regarding the availability of toilet facility for children within the house, nearly one third of the respondents i.e., 50.4 per cent in NGP and 8.4 per cent in non-NGP villages have replied that they have in-house toilet facility in affirmation. Regarding the type of toilet facilities used (respondents who have reported having toilet facility) for their children, the field data reveals that at the aggregate level, a majority of them (91.6%) in non-NGP villages while 37.5 per cent of the respondents in NGP villages practise open defecation and 4.2 per cent of them in non-NGP and 33.9 per cent in NGP use toilets within the house (Table 4.32).

Table 4.32

Details of In-House Toilet Facility and Type of Toilets Used for Children in NGP and Non-NGP Villages: Odisha

Toilet Facility	*NGP*	*Non-NGP*	*Total*
Yes	61	10	71
Per cent	50.4	8.4	29.6
No	60	109	169
Per cent	49.6	91.6	70.4
Type of Toilet Used for Children			
Open defecation	44	109	152
Per cent	37.5	91.6	63.3
Toilet within the household	41	5	46
Per cent	33.9	4.2	19.2
Community/public toilet	5	4	9
Per cent	4.1	3.4	3.8
Others	31	1	32
Per cent	25.6	.8	13.3
Total	121	119	240

Table 4.33

Distribution of Children Taking Help for Toilet Use in NGP and Non-NGP Villages: Odisha

Children Taking Help For Toilet Use	*NGP*	*Non-NGP*	*Total*
By his/her own	3	1	4
Per cent	2.5	.8	1.7
With the help of mother/father	54	44	98
Per cent	44.6	37.0	40.8
With the help of elder sibling/s	25	49	74
Per cent	20.7	41.2	30.8
With the help of others	3	1	4
Per cent	2.5	.8	1.7
Others	6	1	7
Per cent	5.0	.8	2.9
With the help of mother/father or with the help of elder sibling/s	30	23	53
Per cent	24.8	19.3	22.1
Total	121	119	240

Regarding taking help for toilet use, a very few children (1.7%) go on their own. In both the areas, 41 per cent of the children take the help of parents, while one-third of them take the help of siblings. Another one fifth of the children take the help of parents, siblings and other elder members in the family (Table 4.33).

Table 4.34

Awareness about Toilet Use and Children's Vulnerability to Possible Diseases in NGP and Non-NGP Villages: Odisha

Type of Diseases	*NGP*	*Non-NGP*	*Total*
Not aware	9	6	15
Per cent	1.4	2.3	1.6
Vomiting	6	16	22
Per cent	0.91	6.1	2.4
Loose motion	6	16	22
Per cent	0.91	6.1	2.4
Communicable diseases	6	16	22
Per cent	0.91	6.1	2.4
Typhoid	104	29	133
Per cent	15.8	11	14.4
Malaria	104	29	133
Per cent	15.8	11	14.4
Cough	104	29	133
Per cent	15.8	11	14.4
Cold	104	29	133
Per cent	15.8	11	14.4
Diarrhoea	104	29	133
Per cent	15.8	11	14.4
Chikungunya	104	29	133
Per cent	15.8	11	14.4
Dengue	3	18	22
Per cent	0.46	6.8	2.4
Stomach pain	3	18	22
Per cent	0.46	6.8	2.4
Total	657	264	921

With respect to a question on what type of diseases children encounter/experience if they do not use toilets and practise open defecation, three fourths of them have replied that children are prone to typhoid, malaria, diarrhoea, chikungunya etc. A few of the respondents have mentioned that children's health is/was affected by vomiting, stomach ache, loose motion, communicable diseases, dengue, and body pain, whereas an insignificant percentage of the respondents (1.6%) have expressed that they do not have any knowledge about this matter (Table 4.34).

Table 4.35

Distribution of Respondents Reporting Frequency of Bath Given to Children in NGP and Non-NGP Villages: Odisha

Frequency of Bath Given to Children	*NGP*	*Non-NGP*	*Total*
Once a day	111	110	221
Per cent	91.7	92.4	92.1
Twice a day	5	8	13
Per cent	4.1	6.7	5.4
Once in a week	5	1	6
Per cent	4.1	.8	2.5
Total	121	119	240

An analysis of the practice of personal hygiene reveals that a large majority of the respondents (92.1%) give bath to their children every day. Only 5.4 per cent of the respondents both in NGP and non-NGP villages give bath to their children twice a day. However, a few of them (2.5%) give bath to their children once in a week (Table 4.35).

The data reveals that an overwhelming 93 per cent of the children in both the NGP and non-NGP villages, wash their hands before taking their food. Regarding the use of material/method for hand washing, half of the respondents use soap, while the remaining of them wash their hands with water. This method is used mostly in non-NGP villages. This practice is considered unsafe for children's health (Table 4.36).

Table 4.36

Distribution of Children with regard to Washing their Hands and Type of Hand Wash Used before Taking Food in NGP and Non-NGP Villages: Odisha

Children Washing Their Hands	*NGP*	*Non-NGP*	*Total*
Yes	112	111	223
Per cent	92.6	93.3	92.9
No	9	8	17
Per cent	7.4	6.7	7.1
Type of Hand Wash Used			
Water	46	66	112
Per cent	41.1	59.5	50.2
Soap	61	44	105
Per cent	54.5	39.6	47.1
Liquid soap	2	0	2
Per cent	1.8	.0	.9
Detergent soap	0	1	1
Per cent	.0	.9	.4
Others (specify)	3	0	3
Per cent	2.7	.0	1.3
Total	112	111	223

Table 4.37

Distribution of Households by Immunisation Service Received by Children in NGP and Non-NGP Villages: Odisha

Immunisation	*NGP*	*Non-NGP*	*Total*
Received	110	109	219
Per cent	90.9	91.6	91.2
Not received	11	10	21
Per cent	9.1	8.4	8.8
Total	121	119	240

More than 90 per cent of the respondents have mentioned that their children have undergone immunisation, which is a positive measure for keeping good health of children (Table 4.37).

Children's Illness Details in Odisha

In NGP villages, out of 133 children, 74 children are reported to have fallen ill from various water related problems. Among them, one child is reported to have suffered from paralysis which may not be related to water and sanitation. Out of the 74 in NGP, 34 children are reported to have suffered from various illnesses under the category of water borne diseases which includes 13 diarrhoea cases, 7 dysentery cases and 3 diarrhoea cases along with one or more other health problems, while 6 children are reported to have suffered from just cold and cough. 13 cases in NGP villages infected with worms have been reported. Out of them, 4 are reported to have suffered from just worms. One malaria case in NGP area, has also been reported.

In non-NGP villages, out of 136, 47 children are reported to have suffered from various water borne illnesses, while 25 of the children are reported to have suffered from fever/fever with leg pain. There is one viral fever case reported in non-NGP area. There are 20 cases reported in non-NGP villages under water borne diseases. Five (5) of them are reported to have suffered from cold and cough. There are 14 dysentery cases reported in combination with other problems like fever, leg pain, cold cough etc. (Tables 4.38 and 4.39).

Source of Treatment

In all these cases (diarrhoea, dysentery and one or more health problems) in NGP villages, around two thirds of these children are reported to have been taken to government facilities for treatment (SC [sub-centres], PHC [public health centres], CHC [community health centres] and DH [district health]) and the remaining one third to private health facilities for treatment. In NGP areas, most of the children suffering from fever and leg pain and fever are reported to have been taken to government health facilities for treatment. In non-NGP areas, in respect of all these cases (cold and cough, diarrhoea, fever, leg pain, cold cough etc), most of these children 32 out of 50 are reported to have approached private health facilities for treatment. Out of 47, half of them have undergone treatment for 5 to 10 days.

Table 4.38

Number of Children Reported Suffering from Illnesses in NGP Areas: Odisha

	Disease name	Affected	Average Episodes	Source of Treatment						Length of Treatment (days)				Average Expenditure ₹
				SC	PHC	CHC	DH	Private	PHC & Private	> 5	5-10	10-15	15 and Above	
Water borne	Cold and cough	6	1.83	0	2	1	0	3	0	1	2	1	2	816.67
	Cold and cough and dysentery	1	3.00	0	0	0	0	1	0	0	0	0	1	2000.00
	Fever and diarrhoea	1	4.00	0	0	1	0	0	0	0	0	0	1	1600.00
	Diarrhoea and fever and leg pain	1	2.00	0	0	1	0	0	0	0	0	0	1	10000.00
	Dysentery, fever and leg pain	4	3.50	0	1	1	0	1	1	0	0	1	3	2175.00
	Diarrhoea and dysentery	1	2.00	0	0	0	1	0	0	0	1	0	0	1500.00
	Diarrhoea	13	1.69	4	2	1	2	4	0	1	9	1	2	834.62
	Dysentery	7	1.29	0	2	2	0	3	0	0	6	0	1	751.43
Worm infection	(Nulipurugulu)	4	1.25	1	1	0	0	2	0	1	3	0	0	150.00
	Dysentery and nulipurugulu	2	2.50	0	0	0	0	0	2	0	0	1	1	1150.00

	Nulipurugulu, fever and leg pain	4	3.50	1	0	0	0	1	2	0	1	1	2	762.50
	Cold and cough, Nilipurugulu and dysentery	1	4.00	0	0	0	0	1	0	0	0	0	1	4500.00
	Nulipurugulu, dysentery, fever and leg pain	1	5.00	0	0	0	0	1	0	0	0	0	1	1500.00
	Cold and cough and (Nulipurugulu)	1	4.00	0	0	0	0	0	1	0	0	0	1	900.00
Mosquito bite	Fever	7	2.00	1	0	5	0	1	0	1	4	0	2	2071.43
	Body pain	1	2.00	0	0	0	1	0	0	0	1	0	0	700.00
	Malaria	1	1.00	0	0	0	0	1	0	0	0	1	0	1500.00
	Fever and leg pain	17	2.47	0	3	4	0	10	0	2	10	1	4	904.71
	Others	1	3.00	0	0	0	0	1	0	0	0	0	1	2500.00
	Grand total	74	2.24	7	11	16	4	30	6	6	37	7	24	1246.49

Table 4.39

Number of Children Reported Suffering from Illnesses in Non-NGP Villages: Odisha

	Disease Name	Affected	Average Episodes	Source of Treatment						Length of Treatment (Days)				Average Expenditure ₹
				SC	PHC	CHC	DH	Private	PHC & Private	> 5	5-10	10-15	15 and Above	
Water borne	Cold and cough	5	1.80	0	1	0	0	4	0	1	4	0	0	680.00
	Dysentery	4	1.50	0	1	2	0	1	0	0	3	0	1	662.50
	Fever and dysentery	1	3.00	0	0	0	0	0	1	0	0	0	1	600.00
	Dysentery and fever and leg pain	5	3.20	0	0	0	0	5	0	0	1	0	4	1860.00
	Cold and cough and dysentery	4	3.25	0	1	0	0	2	1	1	0	0	3	1625.00
	Typhoid	1	3.00	0	0	0	0	1	0	0	0	0	1	2500.00
Worm infection	Amebiasis	1	1.00	0	0	1	0	0	0	0	1	0	0	500.00
	Nulipurugulu and fever and leg pain	2	3.00	0	0	0	0	2	0	0	0	0	2	1000.00
	Fever and nulipurugulu	2	3.50	0	0	0	0	2	0	0	0	0	2	1250.00
Mosquito bite	Fever	4	1.75	0	1	0	0	3	0	0	3	1	0	1050.00
	Fever and leg pain	17	2.12	1	2	1	2	11	0	2	10	1	4	752.94
	Viral fever	1	1.00	0	0	0	0	1	0	0	1	0	0	2700.00
	Grand total	47	2.36	1	6	4	2	32	2	4	23	2	18	1065.00

Length of Treatment and Expenditure

In NGP area as, in all the cases of diarrhoea, dysentery and other health problems, the average number of episodes comes to 2. A majority of the children are reported to have undergone treatment for more than a week, while the average health expenditure amounts to around ₹1,246.49.

In NGP area, the average number of episodes comes to 1.25 for children with worm infection. The remaining children, along with worm infestation, are reported to have suffered from other problems like fever, dysentery, cold, cough etc. The average number of episodes for these ailments is more than 2. In NGP villages, the health expenditure on malaria amounts to ₹1,500.

In NGP areas, among children suffering from fever and leg pain and fever, many of them have undergone treatment for more than a week. The average health expenditure incurred on fever/fever with leg pain cases for treatment in non-NGP area amounts to around ₹1,000. The average health expenditure in respect of viral fever in non-NGP areas amounts to ₹2,700. In respect of these cases (cold and cough, diarrhoea, fever, leg pain, cold cough etc) in non-NGP areas, the average number of episodes works out to 3. In non-NGP areas for typhoid, 3 episodes have been reported, while expenditure for treatment amounts to ₹2,500.

In NGP areas, a maximum number of children are reported to have suffered from fever and leg pain and diarrhoea, whereas in non-NGP areas, a maximum number of children are reported to have suffering from fever, leg pain, cold and cough. In NGP areas the treatment is reported to have lasted for 5-10 days and in non-NGP areas above 15 days. Children suffering from illnesses go to private clinics for treatment in non-NGP villages and in NGP villages, patients go to public and private hospitals for treatment.

A Profile of Illness Episodes Reported by Households

In this section, we have discussed illness episodes encountered, treatment sought and expenditure incurred by households in NGP and non-NGP areas as follows.

Table 4.40

Illness Episodes Encountered, Treatment Sought and Expenditures Incurred by Households in NGP Areas, During Last One Year: Odisha

	Disease Name	Affected	Average Episodes	Source of Treatment						Length of Treatment (days)				Average Expenditure ₹
				SC	PHC	CHC	DH	Private	PHC & Private	> 5	5-10	10-15	15 and Above	
Water borne	Cold and cough	10	1.80	0	0	4	2	4	0	0	5	2	3	2320.00
	Cold and cough and diarrhoea	3	2.33	0	0	1	1	1	0	0	0	0	3	1766.67
	Dysentery, fever and cold and cough	1	3.00	1	0	0	0	0	0	1	0	0	0	1500.00
	Cold and cough and dysentery	2	2.50	0	0	0	0	0	2	0	0	0	2	2750.00
	Diarrhoea	15	2.00	5	2	0	6	2	0	4	5	1	5	5723.08
	Dysentery	7	1.57	0	2	3	0	2	0	2	3	0	2	985.71
	Digestion problem	2	1.00	0	0	0	0	2	0	0	0	0	2	12000.00
Worm infection	Amebiasis	5	2.00	0	1	1	1	2	0	0	2	0	3	660.00
Mosquito bite	Fever	8	1.38	1	0	7	0	0	0	2	4	1	1	1585.71

Fever and cold and cough	2	3.00	0	1	1	0	0	0	1	0	0	1	3100.00
Fever and chikungunya	2	1.50	1	0	1	0	0	0	2	0	0	0	8000.00
Knee pain and stomach pain, fever and leg pain	1	2.00	0	0	0	0	1	0	0	1	0	0	500.00
Stomach pain	1	1.00	0	0	0	0	1	0	0	0	0	1	50000.00
Gastric problem	3	2.67	0	1	2	0	0	0	0	0	0	3	12166.67
Others	10												
Total	72	1.83	8	7	20	10	15	2	12	20	4	26	8683.09

Table 4.41

Illness Episodes Encountered, Treatment Sought and Expenditures Incurred by Households in Non-NGP Villages: Odisha

	Disease Name	Affected	Average Episodes	Source of Treatment				Length of Treatment (days)				Average Expenditure ₹
				SC	PHC	CHC	Private	> 5	5-10	10-15	15 and Above	
Waterborne	Cold and cough	26	1.23	2	4	14	6	1	18	2	5	2222.00
	Jaundice	1	2.00	0	0	0	1	0	1	0	0	20000.00
	Dysentery	3	1.00	0	0	1	2	0	2	0	1	2433.33
Worm	Amebiasis	1	1.00	0	0	1	0	0	1	0	0	500.00
Mosquito bite	Fever	7	1.29	1	0	1	5	1	3	1	2	1000.00
	Fever and chikungunya	1	1.00	0	0	1	0	0	1	0	0	8000.00
	Fever and cold and cough	4	1.25	0	6	2	2	0	2	0	2	3050.00
	Stomach operation	1	2.00	0	0	1	0	1	0	0	0	20000.00
	Others	5	-	-	-	-	-	-	-	-	-	-
	Total	49	1.22	3	6	21	19	4	30	3	12	6198.94

Illness Episodes

In NGP areas, out of 520 respondents, 72 have reported experiencing water related illnesses during last one year.

However, some illnesses reported in the case of 10 persons may not be directly related to water (cancer, kidney problem, heart problem etc). Under the category of waterborne diseases, in NGP areas, 40 persons are reported to have experienced illnesses like cold, cough, diarrhoea etc., and out of 40, 22 persons are reported to have suffered from diarrhoea/dysentery. In NGP villages, 2 people are reported to have been affected by chikungunya and fever.

The health expenditure is found high in the following cases of illnesses: stomach pain (1-expenditure ₹50,000), gastric problem (3-average expenditure ₹12,166).

The length of treatment for Jaundice in non-NGP areas is reported to be lasting for 5-10 days, while the average expenditure incurred on jaundice treatment amounts to ₹20,000. The average health expenditure incurred on cold and cough in non-NGP areas works out to ₹2,222. In most of the cases in non-NGP villages, the length of treatment is more than a week.

In NGP areas, more number of water related cases in respect of adults have been reported as compared to non-NGP areas. In NGP areas, a maximum number of adults have been found suffering from cold, cough and diarrhoea, whereas in non-NGP areas, a maximum number of adults have been observed suffering from cold, cough and fever. In both NGP and non-NGP areas, treatment days range from 5-10 days. Further, adults suffering from illnesses tend to approach CHC and private clinics in non-NGP villages, while in NGP villages, patients tend to visit public and private hospitals for treatment (Table 4.40 and 4.41).

Summary of Focus Group Discussions (FGDs) Conducted in Both the NGP and Non-NGP Villages in Odisha

Main Sources of Water

- In Mangalpur, Mukundapur, KP Lakharaj, Jamakunda, Rajpur and Naikudi NGP villages, there is a piped water supply through public taps/tap connections to some individual houses, while the main source of water in Badhikapudi, Mushhakhar, Rajanagar, Kendejposi, Polanghats, Chinmaliposi non-NGP villages, is tube wells/bore wells. There is no piped water supply.

Adequacy of Water Supply

- In respect of Mangalpur, Mukundapur, KP Lakharaj, Jamakunda, Rajpur and Naikudi NGP villages, there is an adequate water supply i.e., water supply is adequate for ten months in a year. During the rainy season, there is a problem of contamination of water due to floods. During summer, the ground water level goes down and an erratic power supply also affects the water supply. On the other side, in non-NGP villages, such as Badhikapudi, Mushhakhar, Rajanagar, Kendejposi, Polanghats and Chinmaliposi villages, water availability is inadequate and especially during summer, the water level goes down. A proper maintenance of tube wells is also a challenge and hence, the villagers depend on open well/dug well for drinking water.

Quality of Water

- In Mangalpur and Mukundapur NGP villages, drinking water gets contaminated with dust and fluoride during the rainy season. Sometimes, contaminated water flows through the piped water supply as there is a stone mine near the villages. Traditionally, people boil water before using in KP Lakharaj, Jamakunda, Rajpur and Naikudi villages.
- In all non-NGP and NGP villages, water gets contaminated during the rainy season, while iron dust is found in water during summer.

Water Costs

- In Mangalpur NGP village, each household pays ₹50 per month, whereas from ST community, each household pays ₹30 per month as water charges. However, some households do not pay water charges on time. In Mukundapur village, every household pays ₹50 per month, whereas in KP Lakharaj, Jamakunda, Rajpur and Naikudi villages, people pay ₹30 per month. The O&M (operation and maintenance) in Mangalpur and Mukundapur villages is done by one SEM (self-employed mechanic) and one machine operator and both take care of the maintenance work. In KP Lakharaj, Jamakunda, Rajpur and Naikudi villages, GPs, with the help of the local water and sanitation department, maintain water supply. No money/fund is available with the local *panchayat* office and the *panchayat* only submit the requirements to the water and sanitation department for repair and maintenance. SEMs are appointed to look after the repair and maintenance work.

- In Badhikapudi, Mushhakhar, Rajanagar, Kendejposi, Polanghats, Chinmaliposi non-NGP villages, water charges are collected. The operation and maintenance of tube wells is done by the village GPs. However, they face fund constraints in managing and maintaining tube wells during summer.

Health Problems

- In all NGP villages, unsafe drinking water tends to cause dysentery, fever, diarrhoea and malaria.
- In all non-NGP villages, people tend to suffer from dysentery, fever, diarrhoea etc. During summer, dysentery is common.

Sanitation

- In Mangalpur and Jamakunda NGP villages, most of the houses don't have toilet facility, particularly the BPL (below poverty line) families, while around 25 per cent of the households have toilet facility. In Mukundapur village, a majority of the households have toilet facility, but due to the low quality of construction work, 30 to 40 per cent of the toilets have become defunct. The remaining households have toilet facility. In K.P. Lakharaj, Naikudi and Rajpur villages, only a few households have toilet facility (30%).

- In all non-NGP villages, a maximum number of people do not have toilets (90%), while only around 10 per cent of the households have toilet facility.

Open Defecation

- In all NGP villages, 25 to 30 per cent of the people practice open defecation.
- In all non-NGP villages, open defecation is a common practice among a majority of the people (70% to 80%).

Problems Associated with Open Defecation

- In Mangalpur village, villagers opine that open defecation causes diarrhoea. Besides, in this village, inadequate water supply is a problem when it comes to using of toilets. During rainy season, toilets pits get filled up with water. In Mukundapur, K.P. Lakharaj, Jamakunda, Rajpur and Naikudi villages, people tend to suffer from dysentery, diarrhoea, fever, cold, cough, typhoid, dengue and itching as a result of not using toilets.
- In all non-NGP villages, FGD participants have opined that they may get dysentery, diarrhoea, fever, cold, cough and typhoid etc.

Functioning of Committees

- In Mangalpur (NGP villages), people are aware of the VWSC. In Mukundapur, KP Lakharaj villages, these committees are found working. In Jamakunda village, they have a VWSC known as 'Jamakunda Village Water and Sanitation Committee' (Jamakunda Gramya Jala Parimala Samiti). In Rajpur village, they have a VWSC known as 'Rajpur Village Water and Sanitation Committee' (Rajpur Gramya Jala o Parimala Samiti) managed by a NGO (non-governmental organisation), namely, LIFE. There are 12 members (6 males and 6 females) on the committee. The earlier committee structure has been modified as per the guideline of Rural Water Supply and Sanitation (RWSS). User Groups have been formed under the committee. In Naikudi village, they have a VWSC known as 'Naikudi Village Water and Sanitation Committee' (Naikudi Gramya Jala o Parimala samiti).

- There are no VWSCs formed in all non-NGP villages. However, at the GP level, there is a committee.
- VHSC has been formed in all NGP villages and the committee maintains health and sanitation conditions of the respective village. In all NGP villages, the recent activities taken up by VHSC are road and drain cleaning and creation of health awareness among people. In Mangalpur village, the recent activities include distributing of essential medicines during floods, water purification and cleaning of roads.
- In non-NGP villages, the committees maintain health and sanitation conditions of the villages. In all non-NGP villages, the recent activities undertaken by VHSCs are road cleaning and water purification.

Suggestions for Improving the Performance of VWSCs and VHSCs

- In all NGP villages, regular meetings of the VHSC should be held related to health, water and sanitation matters.
- In all non-NGP villages, VHSC should play a more pro-active role during floods. VHSCs should conduct regular meetings.

Summary

As discussed in the introduction chapter, good health outcomes depend on different interventions: income related interventions, direct health interventions and other institutional interventions. When the health programmes are formulated and implemented effectively, morbidity levels in the villages may come down and there may be reduced episodes of diarrhoea and an improvement in the better nutrition absorption among children etc. However, the health outcomes largely depend on social determinants, considered in this study, such as the availability of quantity and quality water, and sanitation facilities (in this case lavatories and their proper use by households), allied household level hygiene practices conducive to promoting health, personal hygiene such as defecation habits.

This kind of enabling environment is made possible indirectly through a catalytic process by the governing institutions, processes and different programmes at the local level. The study is basically explanatory and analytical in nature.

In Odisha state, 6 NGP and 6 non-NGP villages were selected for the study. From these 12 villages, altogether, data was collected from 240 households. In this study, an attempt is made to examine health conditions of people as water and sanitation programmes are being implemented in the study villages. A majority of the respondents are literates. The educational levels of the respondents in NGP villages are relatively higher as compared to non-NGP villages.

Housing: A majority of the respondents reside in *kutcha* houses and huts and most of the respondents belonging to non-NGP villages live mostly in *kutcha* houses and huts, indicating their poor economic conditions.

Sanitation: A majority of the respondents practice open defection and use pit latrines, especially the respondents belonging to Non-NGP villages. The respondents, having personal toilets (more than half of them) use them to have privacy and another less than a quarter for maintaining good health and privacy. More than half the respondents use sand and another 41.3 per cent of them use soap for washing hands after defecation.

Water: An overwhelming number of the respondents use water for domestic use such as washing dishes, clothes and for bath etc, and a negligible percentage of the respondents use water for toilet purpose. As regard the quality of water, according to three fourths of the respondents water is pure. But as per some of the respondents (5) water is impure, especially during the rainy season.

The analysis reveals that there is a variation in the availability of water between NGP and non-NGP villages and in the former category of villages, adequate water is available as compared to the latter category of villages. When probed further about the availability of adequate quantity of water during the past one year, a great majority of the respondents have reported that it is not adequate. This response has come more from the respondents representing non-NGP villages.

Nearly three fourths of the respondents possess water storage facility/ies at home up to 10-15 tubs. The data reveals that a little over one fifth and less than 10 per cent of the respondents use glass tumblers attached with long handles and taps attached to containers for consuming water. An overwhelming percentage of the respondents clean water containers everyday. Those who clean containers everyday are more in number in non-NGP villages, as compared to NGP villages. A less percentage of the respondents, boil water or strain using a cloth before drinking. This indicates that the educational levels of the household members play a crucial role in treating water before consumption. As per nearly two thirds of the respondents, water shortage is the main problem. More than half of the respondents have mentioned that Anganwadi centres supply water to children who go there.

Water supply in schools: In NGP villages, out of six schools, three get drinking water from bore wells, two schools from taps and one from other sources. Out of six schools in non-NGP villages, three schools get drinking water from bore wells, and one each from taps, tanks and other sources.

Sanitation facilities in schools: With regard to latrine facilities available in schools, in NGP villages, out of six schools, four have septic tank type of toilets being used by students. Another two schools have pit latrines, whereas in non-NGP villages, out of six schools, four schools have septic type of toilets and one each have pit latrines and latrines with a piped sewer system.

Out of six schools in NGP areas, three schools have separate toilets for boys and girls, whereas in non-NGP areas, in four schools out of six, there is such facility. In NGP areas, only three schools have separate toilets for teachers. Whereas in non-NGP areas, all six schools do not have separate toilets for teachers.

Water Facility in ICDS Centres in the Study Villages: There are six ICDS centres in NGP and six in non-NGP villages. In NGP villages, bore well is the main source of water for four ICDS centres followed by one each from taps and wells. In non-NGP villages, out of six ICDS centres, four get water from 'other sources' and a bore well in the case of one.

Hygienic practices: As regards the disposal of human waste (stools) of children, two thirds of the respondents throw it on the streets. As a regards the availability of hand wash facility near the toilets, three fourths of the respondents do not have any facility to wash their hands near toilets nor do they wash their hands. This situation is more visible in non-NGP villages as compared to NGP villages.

The analysis reveals that more than half of the respondents clean their toilets once a month. An overwhelming percentage of the respondents provide water to their children after boiling or straining using a cloth. In the case of children not using toilets, according to three fourths of them, children are prone to typhoid, malaria, cough and cold, diarrhoea, chikungunya etc. As per a few of the respondents, their children's health is/was affected by vomiting, stomach ache, loose motion, communicable diseases, dengue, body pain etc.

A large majority of the respondents (92.1%) give bath to their children every day. More than three fourths of the respondents (76.2%) wash their hands both before and after taking food and after defecation. These trends indicate that they take appropriate measures for protecting their children's health.

Institutions: Here an attempt is made to understand the role of institutions like VWSC, VHSC, SHGs (self help group), CBOs (community based organisations) etc. and processes adopted for achieving ODF villages and provision of clean drinking water to all. An attempt is also made to understand the role of institutions and processes that hinder the achieving of the ODF status of villages and supply of drinking water.

With regard to the level of awareness among people of the existence or constitution of VHSCs in the villages, two thirds of the villagers are aware of these committees, while the remaining one third of them are not aware of such committees. Although VWSCs have been constituted in some villages, their functioning is very discouraging as the members are not aware of their duties and powers. This is true in both the NGP and non-NGP villages.

Illness among Adults: In Odisha, 7 cases in NGP villages and 3 cases of dysentery in non-NGP villages have been reported. In NGP villages, 15 cases of diarrhoea have been reported, but in non-NGP villages no such cases have been reported. In non-NGP areas, 37 adults are reported to have suffered from cough, cold and fever (multiple health problems) in comparison to 20 adults in NGP areas. A majority of them are reported to have approached to CHC (community health centres) and private clinics for treatment.

Illness among Children: In Odisha, in NGP villages, 74 children are reported to have fallen ill from various water related problems. Among 74, 34 children are said to have suffered from various illnesses under the category of waterborne diseases including 13 diarrhoea cases. However, in non-NGP villages, not a single case of diarrhoea has been reported. 7 cases in NGP villages and only 4 cases of dysentery in non-NGP villages have been reported. Only 1 case of malaria in NGP villages has been reported, while in non-NGP villages, not a single case has been reported. In non-NGP areas, 9 children are reported to have suffered from cough, cold and fever (multiple health problems) while 13 children in NGP areas.

Interestingly, NGP areas reported more health problems. This may be due to the difference in living and environmental conditions. In NGP areas, many houses have their own ponds (small) in front of the house and are used for water for domestic purposes and fishing. These villages with many water logged areas happen to be the breeding grounds for mosquitoes and insects. Those with toilets of their own tend to face problems such as foul smell, super structure not ensuring privacy, pits filled up, mosquitoes etc. Regarding healthcare, many of the respondents' prefer to avail treatment from government health facilities. Although there are a few positive aspects like better literacy rates in the study areas, there is not much of a positive impact of water and sanitation programmes or literacy levels on the health of families/communities.

5

Water and Sanitation Programmes and their Effects on the Health Status of Communities in Andhra Pradesh

Introduction

Andhra Pradesh (AP) is one of the most highly populated states in India mainly due to a high level of development and its location near the sea coast. The state has grown in terms of its technological infrastructure and is among the major states that house sectors like IT and telecom. The state is spread over an approximate area of 2,75,000 sq km. The capital city which is also the largest city in the state of AP happens to be Hyderabad. In total, AP state comprises 23 districts.

A Basic Demographic Profile of Andhra Pradesh

As per Census 2011, AP is home to a population of 8.46 crore, as against 7.62 crore as per 2001 Census. In absolute numbers, the total population of AP as per 2011 Census, stands at 84,580,777 of which males and females are 42,442,146 and 42,138,631 respectively. In 2001, the total population was 76,210,007 of which males were 38,527,413 while females were 37,682,594.

AP has a population density of 308/km^2 which is below the national average and thus, the population is spread well over the entire area of the state. According to AP Census 2011, the overall literacy rate in the state is about 67 per cent and is a cause for concern. The overall literacy rate in the state has gone up in recent years, but is still below the national average of about 74 per cent, while the sex ratio is way above the national average at about 993/1000 males.

Literacy Rate 2011

The overall literacy rate in AP has seen an upward trend at 67.02 per cent as per 2011 population Census. Of that, male literacy stands at 74.88 per cent, while female literacy at 58.68 per cent. Comparatively, in 2001, the overall literacy rate in AP was 60.47 per cent of which male literacy was 71.16 per cent and female literacy was 50.29 per cent.

In actual numbers, the total literates in AP come to 50,556,760 of which males were 28,251,243 and females 22,305,517.

Sex Ratio

The sex ratio in AP is 993 per 1000 males, which is below the national average of 940 as per Census 2011. In 2001, the sex ratio was 978 per 1000 males.

Table 5.1

Demographic Profile of AP

State	*Andhra Pradesh*					
	2001			*2011*		
	Male	*Female*	*Total*	*Male*	*Female*	*Total*
Population	38,527,413	37,682,594	76,210,007	42,138,631	42,442,146	84,580,777
Literacy rate (%)	71.16	50.29		74.88	58.68	

Situation in the Study Districts and Villages

As per Census 2011, the percentage of households connected to tap water from treated sources in Visakhapatnam district (both rural and urban) accounts to 48.2 per cent, while 8 per cent of the households are connected to untreated source; 7 per cent depend on uncovered wells, 15.5 per cent on bore wells, and 13 per cent on tube wells/bore wells.

In the case of rural areas of Visakhapatnam, only 31.6 per cent households are connected to tap water from treated sources, while 11.5 per cent households are connected to untreated tap water; 11.4 per cent depend on uncovered wells, 24 per cent on hand pumps,

8.4 per cent depend on tube wells/bore holes and 10.5 per cent on springs (Census 2011).

In G. Madugula mandal, the households connected to tap water from treated sources account to a mere 1.5 per cent, 3.6 per cent of the households are connected to untreated tap water; 18 per cent depend on uncovered wells, 1.2 per cent on hand pumps, 2 per cent on tube wells/bore holes and 70 per cent on springs and 3 per cent on rivers/canals (Census 2011).

Table 5.2

Situation of Drinking Water and Sanitation in AP

Water			
Drinking Water	*Urban*	*Rural*	*Total*
Within the household premises	4,602,652	4,486,906	9,089,558
Near the premises	1,475,858	6,358,607	7,834,465
Away	699,715	3,400,796	4,100,511
Total	6,778,225	14,246,309	21,024,534

Source: Census, 2011.

In Chodavaram mandal, households that are connected to tap water from treated sources account for 60 per cent, while 15.6 per cent of the households are connected to untreated tap water; 1.6 per cent of the households depend on uncovered wells, 16.6 per cent on hand pumps, 6.2 per cent on tube wells/bore holes and only 0.6 per cent on springs (Census 2011).

In Butchayyapeta mandal, households that are connected to tap water from treated sources constitute 33 per cent, while 9 per cent of the households are connected to untreated tap water; 2.8 per cent of the households depend on uncovered well, 44.4 per cent on hand pumps, 8.8 per cent on tube wells/bore holes and only 0.1 per cent on springs (Census 2011).

In Kotauratla mandal, households that are connected to tap water from treated sources accounts for 44.3 per cent, whereas 15.3 per cent of the households are connected to untreated tap water; 8.7 per cent of the households depend on uncovered well, 24.6 per cent on hand pumps and 6.6 per cent on tube wells/bore holes (Census 2011).

In Yelamanchili mandal, households that are connected to tap water from treated sources account for 60.7 per cent, while 6.4 per cent of the households are connected to untreated tap water; 5.8 per cent of the households depend on uncovered wells and 24.8 per cent on hand pumps (Census 2011).

Water and Sanitation

Sources of Drinking Water

In AP, about 78 per cent of the rural population has access to piped water for meeting drinking and domestic water needs. Others use water from hand pumps (fitted with tube wells or open wells), while a small section of the population uses village tanks and springs. There is a high dependence on groundwater for drinking and other domestic purposes. In most of the villages, water is distributed through PSPs (public stand posts). During the year 2008, about 60 per cent of the water supply was ground water based, 38 per cent surface sources based and 2 per cent other sources based like rainwater storages, etc. Further, as on 1 April 2009, about 1,097 habitations were without any safe source, out of which 979 were fluoride affected habitations and 118 salinity affected habitations. Apart from providing treated surface water to fluoride/salinity affected habitations, Rural Water Supply Schemes (RWSS) is also implementing latest technologies for defluoridation/desalination of ground water/excess TDS (total dissolved solids) surface water such as reverse osmosis technique through treatment plants for supplying safe drinking water to the rural people (CESS 2012).

In AP, almost 80 per cent of the habitations are served through Single Village Scheme (SVS) and Mini Piped Water Supply Schemes. SVS is the most preferred option for the RWSS department, provided the source is sustainable in terms of implementation and management. Multi Village Schemes (MVS) are predominantly dependent on surface water source. In terms of habitation coverage, out of a total of 72,147 habitations, 80 per cent (57,718 habitations) are covered through SVS and 7,936 habitations (11%) through MVS (Rajeswara rao et al. 2009).

Figure 5.1

Main Source of Water: Overhead Tank in Gompa (Non-NGP) Village: AP

Source: Field Work.

Sanitation

The present level of sanitation coverage in the state is around 50 per cent, while the rural household coverage is about 30 per cent (and only half of the rural households with latrines use them regularly). This implies that still more than 80 per cent of the rural population resorts to open defecation with its associated risks relating to public health and also, pollution of water supply sources. This problem is more acute in densely populated settlements. In addition to the unscientific/ improper disposal of human waste, the sanitation situation in rural areas becomes more acute, as waste water generated by households including cattle sheds flows into open surface drains that often get choked, leading to stagnation of waste water in the lanes and by-lanes; facilities for a safe disposal of enormous amounts of animal, agricultural and household solid waste are absent; construction of dry pits to discharge siltage is done without any consideration to sanitary aspects; discharge of septic tank effluents into open drains creates unhealthy conditions (Rajeswara Rao et al. 2009).

In Chodavaram mandal, 26.2 per cent of the households have in-house latrine facility, while 73.8 per cent of the households go without latrine facility. A description of the type of latrine facility available shows that, 78 per cent of the households have access to septic tanks, 7.4 per cent to piped sewer system and another 2.5

per cent to other systems under flush/pour-flush latrines. Under pit latrine, 11.3 per cent have access to with-slab/ventilated improved pits, 0.4 per cent to without-slab/open pits. Among the households with no latrine facility, 94 per cent practice open defecation and the remaining 6 per cent depend on public latrines (Census 2011).

In Butchayyapeta mandal, 14.8 per cent of the households have in-house latrine facility and 85.2 per cent of the households carry on without latrine facility. A description of the type of latrine facility available in-house, among the households having latrine facility reveals that 71.3 per cent of the households have access to septic tanks, 7.6 per cent to a piped sewer system and another 5.2 per cent to other systems under flush/pour-flush latrines. Under pit latrine, 13.1 per cent of the households have access to with-slab/ventilated improved pits, 1.4 per cent to without-slab/open pits. Among the households with no latrine facility, 97 per cent practise open defecation and the remaining 3 per cent are dependent on public latrines (Census 2011).

In Kotauratla mandal, 27 per cent of the households have access to in-house toilet facility, while 73 per cent of the households are without latrine facility. A description of the type of latrine facility within the households premise among the households having latrine facility shows that, 77.2 per cent have access to septic tanks, 5.7 per cent to piped sewer system and another 2.3 per cent to other systems under flush/pour-flush latrines.

Under pit latrine, 13.8 per cent have access to with-slab ventilated improved pits, while 0.6 per cent to without-slab open pits. Among the households with no in-house latrine facility, 98.6 per cent practice open defecation with the remaining 1.4 per cent being dependent on public latrines (Census 2011).

In Yelamanchili mandal rural areas, 36.7 per cent of the households have access to in-house latrine facility, whereas 63.3 per cent of the households carry on without latrine facility. A description of the type of latrine facility available in-house among the households having latrine facility indicates that 81.2 per cent have access to septic tanks, while 5.2 per cent to piped sewer system and another 2.2 per cent to other systems under flush/pour-flush latrines. Under pit latrine, 10.4 per cent of the households have access to with-slab/ ventilated improved pits, whereas, 0.5 per cent to without-slab/open pits. Among the households having no latrine facility, 98.5 per cent practice open defecation and the remaining 10.5 per cent depend on public latrines (Census 2011).

A Profile of the Study Villages in the District

For a better understanding of the public policies and the governance structures, it is necessary to study the environmental factors like the location, socio-economic structure, infrastructural facilities, civil society bodies, etc, for they wield a considerable influence on the performance of the development programmes. The social development policies such as provision of protected drinking water and sanitation have to operate in the context of the geographical, socio-cultural and political configurations of the society or the areas where they are being operated. The interaction between the environment and administration devised for the implementation of policies is very important, especially at the grass roots levels. In this context, an attempt is made to examine some of the social determinants and infrastructural facilities available in the villages selected for the study. As mentioned earlier, the study was conducted across 12 villages (6 Nirmal Gram Puraskar [NGP] and 6 non-NGP villages) in the district, Visakhapatnam.

Location and Demographic Features of the Study Villages

In NGP and non-NGP areas, the number of villages selected for the study comes to 12—6 in each area. Transport in NGP villages is covered by both RTC and auto services to the extent of 50 per cent and another one sixth of the needs are met by auto services. But in non-NGP villages, two third of the transport needs are met through/ by RTC and auto services. All the villages in both NGP and non-NGP areas have electricity facility.

Sources of Drinking Water in the Study Villages

In all the villages in both NGP and non-NGP areas, drinking water supply is made available through bore wells with overhead tanks. For meeting additional water requirements, people depend on lakes/tanks in NGP villages and protected wells in non-NGP villages. Water tanks in the villages are cleaned weekly in one village in NGP and two of the villages in non-NGP. In half of the villages of NGP, water tanks are cleaned fortnightly; whereas in non-NGP villages, it is only done in two villages.

Village Water and Sanitation Committees (VWSCs)

As per the National Health Mission (NHM)/programme norms, in every village, there shall be a VWSC to guide and monitor the water and sanitation services at the grass root level. In AP, out of 6 NGP villages, all 6 villages' have VHSCs (village health and sanitation committee) formed and 4 villages have VWSCs formed. In non-NGP villages, all 6 villages have VHSCs formed and only in 2 villages VWSCs have been formed. Thus, in both type of villages, there are VWSCs. They conduct meetings in 4 villages in NGP areas and 2 villages in non-NGP areas. As regards the frequency of meetings, out of 4 VWSCs in NGP, 2 villages conduct quarterly meetings and in the remaining villages either monthly (or) half yearly meetings are held. In non-NGP, out of two villages conducting meetings, one holds monthly and the other holds half yearly meeting. Thus, at the aggregate level (i.e., NGP and non-NGP villages taken together) monthly, quarterly and half-yearly meetings are held in 2 villages each. Only 41.7 per cent of the village committees maintain records/minutes of meetings. In NGP areas, all the 4 villages that conduct meetings, maintain records, whereas in non-NGP villages, out of 2 villages, only 1 village maintains records.

In NGP areas, schools and *panchayats* are equally important as venue (33.3% each), for conducting meetings, whereas, in non-NGP areas, schools are mostly prefered for conducting meetings (50%) followed by panchayat offices. As regards the frequency of VHSC meetings, mostly monthly meetings are held in both the NGP (66.7%) and non-NGP (83.3%) villages. Maintenance of records of VHSC meetings is done in all the villages of NGP and non-NGP areas. In NGP villages, records are maintained mostly by the village sarpanch and other committee members concerned (VHSC/VWSC together in 4 villages). But in non-NGP areas, other committee members maintain records in all the villages.

Village level health plans are prepared in half of the villages of NGP category, while in non-NGP category, plans are prepared in 5 villages (83.3%). The VHSC bring the health issues mentioned in the plan documents to the notice of the relevant health functionaries in five villages in NGP category (83.3%) and all the villages in non-NGP category (100%).The respective VHSC presents the Village health plan in the gramsabha of the village panchayat in both NGP and non-NGP villages (4 villages in each area). VHSC members are aware of grants received by VHSCs in all the villages belonging to both the NGP and non-NGP. All VHSC members involve themselves in the maintenance of records in all the villages of NGP and non-NGP. VHSCs do not get any additional financial resources in all the villages of NGP and non-NGP. In all the villages of NGP and non-NGP, people are aware of the existence of VHSCs. IEC (information education and communication) material is made available in the form of posters in half of the villages of NGP, followed by hand-outs/pamphlets in one sixth of the villages.

State of Water and Sanitation Facilities in Government Schools in the Study Villages

In AP, in NGP villages, there are six schools and six Indian Child Development Services (ICDS) centres, where as, in non-NGP villages, there are eight schools and eight ICDS centres. In NGP, out of six schools, two get drinking water from taps, and one each from tanks, cans and public taps. Similarly, out of eight schools in non-NGP, three get drinking water from taps and another three schools from tanks. As regards the availability of latrine facilities in schools, all the six schools in NGP villages, have access to septic latrines.

All the six schools in NGP villages, have separate toilets for boys, girls and teachers, whereas in non-NGP villages, in five out of eight villages, there is such facility available. Regarding cleaning of toilets in schools, in NGP villages, in a majority of schools (four out of six), cleaners are hired, followed by sweepers (in one village). In non-NGP villages, in half of the schools (four out of eight schools), sweepers and hired cleaners are there. Four out of six schools in NGP villages and three out of eight schools in non-NGP villages have their toilets cleaned fortnightly, followed by daily in one school and weekly in one school. Water supply is sufficient in five out of six schools in NGP villages, whereas water is sufficient in four out of eight schools in non-NGP villages.

Figure 5.2

Pucca Houses in Turakalapudi, Non-NGP Village: AP

Source: Field Work.

Water Facility in ICDS Centres in the Study Villages

As regards ICDS centres, there are 5 in NGP villages and 10 in non-NGP villages. In NGP, drinking water in ICDS centres is supplied through taps (60%) followed by bore wells (20%). In non-NGP villages, half of the centres get water from taps, followed by tanks (10%). As regards the supply of water for general use, in NGP villages, 3 centres get water from taps, 1 gets from wells and the other gets through can supply. But in non-NGP villages, only 3 out of 10 centres get water from taps, and do not depend on wells and can supply.

A Profile of Households and Children Behaviour Related to Water, Sanitation and Hygiene

This section is based on the data collected using household and child schedules. In this section, an attempt is made to examine the perceptions of household members representing NGP and non-NGP villages regarding the utilisation of water and sanitation programmes and their impact on the health conditions of the rural communities. Unlike the other studies, the present study adopted the programmes (villages with NGP) and control-group (villages without NGP) approach, covering both NGP and non-NGP village households with a view to juxtaposing the situation and to see the response of the people towards the programmes and their impact on the health conditions of the people.

An analysis of the experiences of those who are recipients of the scheme, NGP provides insights into the operational dynamics of the programme. This also brings out the gap between promise and performance, highlighting the strengths and weaknesses of the programmes. Similarly, an understanding and assessment of the perceptions of the non-NGP households could provide a clue to identifying the causes behind the non-availability of facilities offered by the government. Here, an attempt is made to analyse the socio-economic background of the respondents, representing NGP and non-NGP villages.

Table 5.3

Gender and Literacy Profiles of the Respondents in NGP and Non-NGP Villages: AP

Education	*NGP Villages*			*Non-NGP Villages*			*All*		
	Male	*Female*	*Total*	*Male*	*Female*	*Total*	*Male*	*Female*	*Total*
Primary	16	2	18	21	2	23	37	4	41
Per cent	15.1	13.3	14.9	18.6	20.0	18.7	16.9	16.0	16.8
Secondary (6-12)	34	0	34	32	0	32	66	0	66
Per cent	32.1	0.0	28.1	28.3	0.0	26.0	30.1	0.0	27.0
Graduate	5	0	5	8	0	8	13	0	13
Per cent	4.7	0.0	4.1	7.1	0.0	6.5	5.9	0.0	5.3
Illiterate	40	13	53	42	7	49	82	20	102
Per cent	37.7	86.7	43.8	37.2	70.0	39.8	37.4	80.0	41.8
Literate	11	0	11	10	1	11	21	1	22
Per cent	10.4	0.0	9.1	8.8	10.0	8.9	9.6	4.0	9.0
Total	106	15	121	113	10	123	219	25	244

Table 5.4

Profile of the Households by Type of House in NGP and Non-NGP Villages: AP

Type of House	*NGP Villages*	*Non-NGP Villages*	*Total*
Pucca	101	101	202
Per cent	83.5	82.1	82.8
Semi-*pucca*	12	13	25
Per cent	9.9	10.6	10.2
Kutcha	6	7	13
Per cent	5.0	5.7	5.3
Hut	2	2	4
Per cent	1.7	1.6	1.6
Total	121	123	244

At the aggregate level, a majority of the respondents live in *pucca* houses (82.8%) followed by semi-*pucca* houses (10.2%). In NGP and non-NGP villages also, the proportions remain nearly the same in this respect (Table 5.4).

The gender particulars of the sample respondent households point to a large number of male members (89.8%) out of 244 households covered by the study. At the overall level (NGP and non-NGP), a majority of the persons are illiterate (41.8%), while illiteracy is only 37.4 per cent among males as against 80 per cent among females. With regard to primary level of education, the percentage share is nearly equal among males and females (16-17%), and no female members fall under secondary and above levels of education (Table 5.3).

In non-NGP villages, the illiteracy rate is less among females in comparison to their counterparts in NGP villages. Similarly, in respect of primary level of education also, non-NGP village respondents are a little better off, whereas NGP category respondents show a little better performance with regard to secondary level of education (only males are there in the both groups).

At the aggregate level, a majority of the respondents live in *pucca* houses (82.8%) followed by semi-*pucca* houses (10.2%). In NGP and non-NGP villages also, the proportions remain nearly the same in this respect (Table 5.4).

Table 5.5

Distribution of the Sample Households by Type of Toilet in NGP and Non-NGP Villages: AP

Toilet Type	*NGP Villages*	*Non-NGP Villages*	*Total*
Flush toilet	1	0	1
Per cent	.8	.0	.4
Septic latrine	76	59	135
Per cent	62.8	48.0	55.3
Open defecation	43	61	104
Per cent	35.5	49.6	42.6
Septic tank and OD	1	3	4
Per cent	.8	2.4	1.6
Total	121	123	244

Note: Chi-value=5.157, P=0.076.

The field data shows that at the aggregate level, a majority of the respondents use septic latrines (55.3%), followed by open defecation (42.6%). In non-NGP areas, half of the respondents practice open defecation, while nearly another half of them use septic latrines. In NGP areas, one third of the residents practice open defecation and the remaining two thirds of the respondents use septic latrines. At the aggregate level as well as category-wise, only 1 household uses flush toilet. In the usage of septic latrines, NGP villages are better compared to non-NGP villages (Table 5.5).

At the aggregate level, regarding the reasons for not using toilets, two households in non-NGP have reported that 'super structure does not ensure privacy' and 1 from NGP has responded that the latrine has got filled up (Table 5.6).

Table 5.6

Respondents' Perceptions Related to Non-use of Toilets in NGP and Non-NGP Villages: AP

Reason for Not Using Toilet	*NGP Villages*	*Non-NGP Villages*	*Total*
Superstructure does not ensure privacy	0	2	2
Per cent	.0	100.0	66.7
Latrine filled up	1	0	1
Per cent	100.0	.0	33.3
Total	1	2	3

At the aggregate level, regarding the reasons for using toilets, 97.1 per cent of the respondents use them "for better health", whereas 2.1 per cent of the respondents use toilets for maintaining privacy and better health. In both NGP and non-NGP villages also, the predominant response for using toilets is better health with 96.2 per cent and 98.4 per cent respectively. As regards the educational level, illiteracy is higher among NGP village respondents (61.53%), as compared to those in non-NGP villages (38.4%). Irrespective of the level of education, by and large, majority of respondents in both NGP and non-NGP villages, use toilets for better health (Table 5.7).

At the aggregate level, 57.4 per cent of the respondents wash hands with plain water after defecation followed by 34 per cent of the respondents' who wash their hands with soap and only 8.6 per cent of the respondents with both soap and plain water. Washing hands with soap is slightly higher in respect of NGP villages (37.2%) as compared to non-NGP villages (30.9%). Washing hands with only plain water is higher in non-NGP villages (60.2%) as compared to NGP villages (54.5%). In respect of washing hands with both soap and plain water, the situation in both NGP and non-NGP villages is more or less the same (Table 5.8).

Table 5.7

Respondents' Perceptions Regarding the Use of Toilets, by Level of Education in NGP and Non-NGP Villages: AP

	Reasons for Using Toilets									
	NGP Villages			*Non-NGP Villages*			*All*			
Education	*For Better Health*	*For Better Health and Privacy*	*Total*	*For Better Health*	*Privacy*	*Total*	*For Better Health*	*For Better Health and Privacy*	*Privacy*	*Total*
Primary	13	1	14	14	0	14	27	1	0	28
Per cent	92.85	7.14	50.00	100.00	0.00	50.00	96.42	3.57	0.00	100.00
Secondary (6-12) and graduation	26	1	27	23	0	23	49	1	0	50
Per cent	96.29	3.70	54.00	100.00	0.00	46.00	98.00	2.00	0.00	100.00
Illiterate	31	1	32	19	1	20	50	1	1	52
Per cent	96.87	3.12	61.53	95.00	5.00	38.46	96.15	1.92	1.92	100.00
Literate	5	0	5	5	0	5	10	0	0	10
Per cent	100.00	0.00	50.00	100.00	0.00	50.00	100.00	0.00	0.00	100.00
Total	75	3	78	61	1	62	136	3	1	140

As regards using the occasional source of water, at the aggregate level, out of 244 respondents, a majority (84%) have reported that they use water for domestic use only, while some respondents (14.8%) use it for both domestic and toilet purpose and 1.2 per cent of them use water for toilet purposes only. This trend is seen in both NGP and non-NGP villages. In non-NGP villages, a little higher percentage i.e., 87.8 per cent of them uses water for domestic use only as compared to NGP villages (80.2%). In respect of using water for both domestic and toilet purpose a higher percentage is seen in respect of NGP villages (18.2%) than non-NGP villages (11.4%) (Table 5.9).

Table 5.8

Details of Hand Wash Habit after Defecation in NGP and Non-NGP Villages: AP

Washing Hands After Defecation	*NGP*	*Non-NGP*	*Total*
Soap	45	38	83
Per cent	37.2	30.9	34.0
Plain water	66	74	140
Per cent	54.5	60.2	57.4
Soap and plain water	10	11	21
Per cent	8.3	8.9	8.6
Total	121	123	244

Table 5.9

Water Availability and Usage (by Source) in NGP and Non-NGP Villages: AP

Purpose of Using the Occasional Source of Water	*NGP*	*Non-NGP*	*Total*
Domestic use	97	108	205
Per cent	80.2	87.8	84.0
Domestic use and toilet facility	22	14	36
Per cent	18.2	11.4	14.8
Toilet use	2	1	3
Per cent	1.7	.8	1.2
Total	121	123	244

Figure 5.3

Village Health Committee Minutes Register in Gompa, AP, a Non-NGP Village

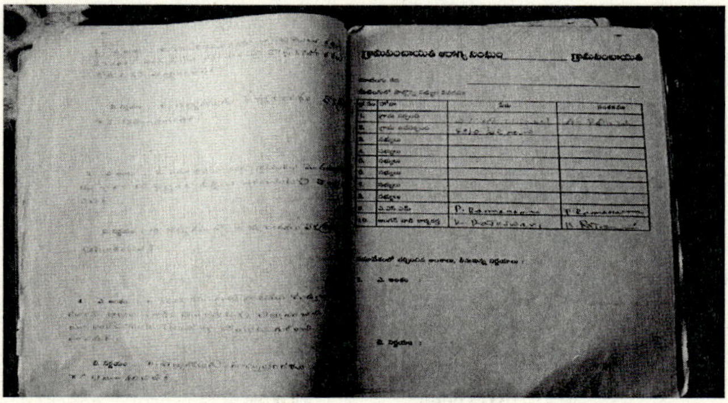

Source: Field Work.

Box 5.1

Hazipalli Village: A Struggle Beyond Development and NGP Award

Hazipalli village, situated on the west side of Shadnagar town, comes under Farooqnagar Mandal of Mahaboobnagar district in Telangana State. The village is heterogeneous and is inhabited by OBC (other backward caste) and SC (schedule caste) communities. The total population of the village stands at 872 and that comprises 242 households of which OBC households come to 172 and SC households to 70. The main economic activity of the village is agriculture and allied activities, while some people are engaged in government jobs and some of the people in business as well. Smt Jangamma (a dalit), a former panchayat president observes that when they came to know about the NGP Award and the criterion to become eligible to get NGP Award, the Panchayat convened a Gram Sabha meeting in which it was decided not to practice open defecation and to maintain good hygiene practices at the household and village levels. To prevent open defecation practice in the village, all the women from SHGs played a vital role besides giving full support to the Panchayat President. The Panchayat President had tried to get a latrine scheme from the sanitation department and she succeeded. All the households utilized this scheme and they constructed individual toilets at their houses and started using them. Smt. Jangamma is happy to say that she passed a resolution in the Gram Sabha with the support of all the villagers to the effect that if any person practiced open defecation, he/she would be punished at the Panchayat Office by keeping them whole day at the office and also imposing a fine of ₹500. To implement this resolution she had undergone lot of struggle, but with the help of SHG women and other men, she succeeded. Mr Srinivas who is the present Panchayat President from the village has given a great support to prevent open defecation in addition to supporting the ex president in all the developmental activities.

As regards the processes followed, Smt Jangamma mentioned that she conducted a massive rally; handouts were distributed; slogans were written on walls and conducted meetings in the village to create awareness regarding open defecation problems and the importance of personal hygiene. The other villagers have endorsed that today their village practises cent per cent ODF and have the facilities of a completely protected water supply and underground drainage system. As result of this huge success, because of the collective efforts of the panchayat president, SHG members and villagers in general, Hazipalli village case study as a ‘Success Case’ has been introduced in the sixth class social studies syllabus by the AP Government. Not only at household level, the village school and ICDS centre are provided with toilets and protected water supply. The committee viz., VHSC functions well. Today Hazipalli village stands as a proud village attracting a number of visitors (both from India and abroad) to have a glimpse of the village achievements in the sanitation and water sector.

Table 5.10

Perceptions of the Respondents Regarding the Quality of Water in NGP and Non-NGP Villages: AP

Quality of Water	*NGP*	*Non-NGP*	*Total*
Pure	120	123	243
Per cent	99.2	100.0	99.6
Others	1	0	1
Per cent	.8	.0	.4
Total	121	123	244

Note: Chi-value=1.021, P=0.312.

According to almost all the respondents in both NGP and non-NGP villages, water is pure. There is not much difference between NGP and non-NGP villages (Table 5.10).

Table 5.11

Details of Adequacy/Inadequacy of Water Availability for Daily Needs: AP

Quantity of Water	*NGP*	*Non-NGP*	*Total*
Fully sufficient	107	101	208
Per cent	88.4	82.1	85.2
Somewhat sufficient	13	21	34
Per cent	10.7	17.1	13.9
Insufficient	1	1	2
Per cent	.8	.8	.8
Total	121	123	244

At the aggregate level, a majority of the respondents (85.2%) have reported that the quantity of water supplied is fully sufficient followed by 13.9 per cent of the respondents who have reported that water availability is somewhat sufficient. In NGP villages, an overwhelming number of respondents (88.4%) have responded that water availability is fully sufficient as against 82.1 per cent of the respondents in non-NGP villages (Table 5.10).

At the aggregate level, most of the respondents (84.8%) have reported that water supply has been sufficient during past one year.

However, in NGP villages, such a positive response is found a tad higher (86.8%) than in Non-NGP villages (82.9%) (Table 5.12).

Table 5.12

Perceptions of the Respondents Regarding Adequacy of Water Supply During the Past One Year in NGP and Non-NGP Villages: AP

Adequate of Water Supply	*NGP*	*Non-NGP*	*Total*
Yes	105	102	207
Per cent	86.8	82.9	84.8
No	16	21	37
Per cent	13.2	17.1	15.2
Total	121	123	244

Table 5.13

Particulars of Season-Wise Water Insufficiency (in terms of Intensity): AP

Seasonal Water Insufficiency	*NGP*	*Non-NGP*	*Total*
In summer season	16	21	37
Per cent	100.0	100.0	100.0
In winter season	0	0	0
Per cent	0.0	0.0	0.0
Total	16	21	37

Table 5.14

Information on the Storage Capacity (Drinking Water) of Water Tanks/ Drums/Vessels in the Sample Households in NGP and Non-NGP Villages: AP

Storage Capacity of Drinking Water	*NGP*	*Non-NGP*	*Total*
15 pots	121	123	244
Per cent	100.0	100.0	100.0
1 drum	0	0	0
Per cent	0.0	0.0	0.0
10 buckets	0	0	0
Per cent	0.0	0.0	0.0
Total	121	123	244

In both NGP and non-NGP areas, 37 (15.2% of the total sample households) respondents have stated that there has been a problem with water supply during the last 1 year. When we asked them to mention the season during which they face water problem, all of them have reported that they face problem during summers (Table 5.13).

In both NGP and non-NGP villages, a majority of the households use earthen pots besides plastic/steel vessels for water storage; they do not use drums for storing water. Almost all the respondents in both NGP and non-NGP villages use pots (about 15) for storing water (see Table 5.14).

Table 5.15

Information on Practices of Drawing Drinking Water from Containers in NGP and Non-NGP Villages: AP

Method of Using to Take Water from Container	*NGP*	*Non-NGP*	*Total*
Containers attached with taps	3	2	5
Per cent	2.5	1.6	2.0
Using tumblers/utensils with hands	118	121	239
Per cent	97.5	98.4	98.0
Total	121	123	244

At the aggregate level, out of 244 respondents, 98 per cent take water from containers with glass/tumblers/utensils using hand, whereas, only 2 per cent of them take water from containers with taps attached. Taking out water with glass/tumblers/utensils using hand is higher in Non-NGP villages (98.4%) than in NGP villages (97.5%), whereas, taking water from containers with taps attached is higher in NGP villages (2.5%) than in non-NGP villages (1.6%) (Table 5.15).

Regarding cleaning of drinking water containers, in both NGP and non-NGP villages, all the 244 respondents do clean drinking water containers. At the aggregate level, 98 per cent of respondents clean containers daily and only 2 per cent respondents once in 2 days. However, in non-NGP villages, 98.4 per cent of the respondents clean water containers daily as against 97.5 per cent in NGP villages. There is not much difference between NGP and non-NGP villages, in respect to cleaning water containers daily (Table 5.16).

Table 5.16

Distribution of Households by Frequency of Cleaning Water Containers Used for Drinking Water in NGP and Non-NGP Villages: AP

Frequency of Cleaning Water Containers	*NGP*	*Non-NGP*	*Total*
No	0	0	0
Per cent	0.0	0.0	0.0
Yes	121	123	244
Per cent	100.0	100.0	100.0
Daily	118	121	239
Per cent	97.5	98.4	98.0
Once in 2 days	3	2	5
Per cent	2.5	1.6	2.0
Total	121	123	244

Note: Chi-value=0.221, P=0.638.

Figure 5.4

Posters Displayed Containing Information on Medicine for Night Blindness and Hand Wash Practices in Rajupeta: AP

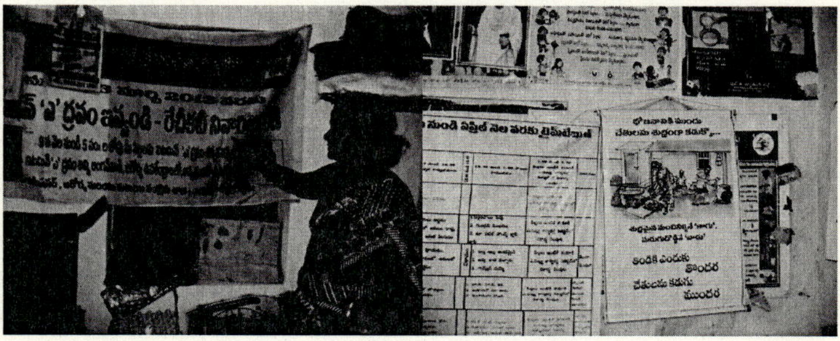

Source: Field Work.

At the aggregate level, the respondents who treated water 'sometimes' before drinking constitute 45.1 per cent, whereas only 6.1 per cent of them treat water always. Interestingly, the respondents who sometimes treat drinking water account for a tad higher share in non-NGP villages (46.3%) in comparison to those in NGP villages (43.8%), while the percentage of respondents treating water always is higher in respect of NGP villages (6.6%) than in non-NGP villages (5.7%). The percentage of respondents who never treat drinking water is the same in both NGP and non-NGP villages (46.3%) (Table 5.17).

Table 5.17

Information on Treatment of Drinking Water Before Consumption, by Level of Education of Respondents in NGP and Non-NGP Villages: AP

Educational Levels	Treatment of Drinking Water														
	NGP					Non-NGP					All				
	Always	Sometimes	Never	Not at All Done	Total	Always	Sometimes	Never	Not at All Done	Total	Always	Sometimes	Never	Not at all Done	Total
Primary	1	12	5	0	18	1	11	10	1	23	2	23	15	1	41
Per cent	5.55	66.66	27.77	0.00	43.90	4.34	47.82	43.47	4.34	56.09	4.87	56.09	36.58	2.43	100.00
Secondary (6-12) and graduate	4	23	12	0	39	3	26	10	1	40	7	49	22	1	79
Per cent	10.25	58.97	30.76	0.00	49.36	7.50	65.00	25.00	2.50	50.63	8.86	62.02	27.84	1.26	100.00
Illiterate	3	14	32	0	53	2	16	31	0	49	5	30	63	4	102
Per cent	5.60	26.41	60.37	0.00	52.47	4.08	32.65	63.26	0.00	48.03	4.90	29.41	61.76	3.92	100.00
Literate	0	4	7	0	11	1	4	6	0	11	1	8	13	0	22
Per cent	0.00	36.36	63.63	0.00	50.00	9.09	36.36	54.54	0.00	50.00	4.54	36.36	59.09	0.00	100.00
Total	8	53	56	4	121	7	57	57	2	123	15	110	113	6	244

Figure 5.5

School Children Collecting Water from Hand Pumps in V.J. Puram: AP

Source: Field Work.

Table 5.18

Details of Measures Taken for Treating Water before Consumption by the Sample Households in NGP and Non-NGP Villages: AP

Measures Taken	*NGP*	*Non-NGP*	*Total*
Boil water	57	57	114
Per cent	93.4	89.1	91.2
Strain water, using a cloth	2	7	9
Per cent	3.3	10.9	7.2
Use water filters	2	0	2
Per cent	3.3	.0	1.6
Total	61	64	125

Note: Chi-value=4.708, P=0.095.

At the aggregate level, out of 125 respondents, 91.2 per cent boil water so as to make it safer for drinking followed by 7.2 per cent who strain water, using a cloth, while 1.6 per cent use water filters. In NGP villages, respondents boiling water before drinking constitute a higher share (93.4%) than those in non-NGP villages (89.1%) as it is evidence by Chi-square test. The percentage of respondents straining water, using a cloth is higher in respect of (10.9%) non-NGP villages relative to NGP villages (3.3%). And only 3.3 per cent of the respondents in NGP villages use water filters (Table 5.18).

Table 5.19

Information on the Major Problems Related to Water Supply in the Sample Villages (NGP and Non-NGP): AP

Major Problem	*NGP*	*Non-NGP*	*Total*
No problem	101	100	201
Per cent	83.5	81.3	82.4
Bore-wells not available	3	3	6
Per cent	2.5	2.4	2.5
Shortage of water	10	13	23
Per cent	8.3	10.6	9.4
Long distance involved in fetching water	5	1	6
Per cent	4.1	.8%	2.5
Bore wells not available and shortage of water	1	2	3
Per cent	.8	1.6	1.2
Shortage of water and long distance involved in fetching water	1	4	5
Per cent	.8	3.3	2.0
Total	121	123	244

Table 5.20

Distribution of the Respondents Reporting Major Problems Regarding Water Supply in NGP and Non-NGP Villages: AP

Problems	*NGP*	*Non-NGP*	*Total*
Water level goes down during summer	17	19	36
Per cent	85.0	82.6	83.7
Water level goes down during summer & no other source available for drinking water	3	4	7
Per cent	15.0	17.4	16.3
Total	20	23	43

With regard to water supply, according to a majority of the respondents (82.4%), there is no problem. The rest of them have reported facing various problems regarding water supply. The respondents reporting water shortage in non-NGP village's account for 10.6 per cent, and those reporting the involvement of long distances in fetching water for 3.3 per cent and those reporting the non-availability of bore wells in their locality for 2.4 per cent. In NGP areas also, the respondents reporting water shortage account for 8.3 per cent, while those reporting the involvement of long distances in fetching water for 4.1 per cent (Table 5.19).

At the aggregate level, 43 respondents (from both NGP and non-NGP areas) face drinking water problem, especially during summer mainly because the water level goes down and hence, they do not get sufficient water from wells/bore wells (Table 5.20).

Table 5.21

Status of Water Supply in ICDS Centres in NGP and Non-NGP Villages: AP

Water Supply in ICDS Centre	*NGP*	*Non-NGP*	*Total*
Yes	54	51	105
Per cent	44.6	41.5	43.0
Others	67	72	139
Per cent	55.4	58.5	57.0
Total	121	123	244

At the aggregate level, two thirds of the respondents have reported that there is water supply in Anganwadi centres. In respect of NGP villages, 44.6 per cent of the respondents have reported that ICDS centres have access to water supply, while in non-NGP only 41.5 per cent of respondents have reported the availability of water in ICDS centres (Table 5.21).

Regarding the advantages of using toilets within the house, at the aggregate level, more than half of the respondents (59.3%) have reported that they use toilets for 'better health', followed by 27.9 per cent of the respondents for 'better health and privacy.' In respect of NGP villages, 56.4 per cent of the respondents use toilets for 'better health' as against 62.9 per cent of the respondents using toilets for the same reason in non-NGP villages. 'Better health and privacy' receives a little higher response 29.5 per cent in NGP villages than in non-NGP villages (25.8%). In both NGP and non-NGP villages together, 'better health and social status' has received only 5 per cent response (Table 5.22).

Table 5.22

Respondents' Awareness Regarding the Advantages of Using In-House Latrine/Toilet/Lavatory Facility in NGP and Non-NGP Villages: AP

Advantages	*NGP*	*Non-NGP*	*Total*
Better health	44	39	83
Per cent	56.4	62.9	59.3
Privacy	4	2	6
Per cent	5.1	3.2	4.3
Ease of use	2	0	2
Per cent	2.6	.0	1.4
Social status	1	0	1
Per cent	1.3	.0	.7
Better health and privacy	23	16	39
Per cent	29.5	25.8	27.9
Better health and ease of use	0	1	1
Per cent	.0	1.6	.7
Better health and social status	4	3	7
Per cent	5.1	4.8	5.0
Privacy and ease of use	0	1	1
Per cent	.0	1.6	.7
Total	78	62	140

Table 5.23

Respondents Encountered Problems by the Households Relating to Toilet Use in NGP and Non-NGP Villages: AP

Problems	*NGP*	*Non-NGP*	*Total*
Yes	3	3	6
Per cent	3.8	4.8	4.3
No	75	59	134
Per cent	96.2	95.2	95.7
Total	78	62	140

Regarding toilet-use problems at home, at the aggregate level, 95.7 per cent of responds do not face any problem. There is only a 1 per cent difference in this respect between NGP and non-NGP villages (96.2% in NGP villages) (Table 5.23).

Table 5.24

Problems Encountered by Households Relating to Toilet Use in NGP and Non-NGP Villages: AP

Problems	*NGP*	*Non-NGP*	*Total*
Water not available for use/cleaning	2	1	3
Per cent	66.6	33.3	50.0
Superstructure does not ensure privacy	0	1	1
Per cent	0.0	33.3	16.6
Filled up Pits	1	0	1
Per cent	33.3	0.0	16.6
Total	3	3	6

With regard to problems faced by respondents related to toilets use at home, one respondent has reported non-availability of water and super structure not ensuring privacy as the reasons for not using toilet facility and one respondent because of unstable slab in non-NGP villages At the aggregate level, out of six respondents both in NGP and in non-NGP villages, three respondents have reported 'water not available for use/cleaning', while one respondents "filled up pit" in NGP villages (Table 5.24).

Table 5.25

Information on the Practices of Disposing Children's Stools Post Defection in NGP and Non-NGP Villages: AP

Method of Disposal of Children's Excreta	*NGP*	*Non-NGP*	*Total*
Leave it where it is	8	9	17
Per cent	6.6	7.3	7.0
Throw it on the street	84	96	180
Per cent	69.4	78.0	73.8
Throw it in the latrine	29	18	47
Per cent	24.0	14.6	19.3
Total	121	123	244

Note: Chi-value=3.417, P=0.121.

With regard to disposing of children's stools, at the aggregate level, a majority of the respondents (73.8%) 'throw on the street', followed by 19.3 per cent of respondents 'throw it in the latrine' and another 7 per cent of the respondents ‘leave it where it is.’ In NGP villages, 24 per cent of the respondents dispose of children’s excreta in the latrine as against 14.6 per cent of the respondents in non-NGP villages. As against this, 78.10 per cent of the respondents in non-NGP villages dispose of children’s waste on to the street as compared to 69.4 per cent of the respondents in NGP villages (Table 5.25).

Table 5.26

Information on Hand Wash Facility and Type of Hand Wash Near Toilets: AP

Type of Hand Wash	*NGP*	*Non-NGP*	*Total*
Not having wash facility	1	1	2
Per cent	1.3	1.6	1.4
Having wash facility	77	61	138
Per cent	98.7	98.4	98.6
Soap	54	43	97
Per cent	70.1	70.5	70.3
Plain water	8	4	12
Per cent	10.4	6.6	8.7
Soap and water	15	14	29
Per cent	19.5	23.0	21.0
Total	77	61	138

Figure 5.6

School Having Toilets for Boys and Girls in Rajupeta: AP

Source: Field Work.

Regarding hand wash facilities near the toilets, at overall level, 98.6 per cent of the respondents have reported that they have access to hand wash facility near toilets at home. However, the respondents having this are more in number in NGP villages. The number of respondents having hand wash facility near toilets is marginally higher in NGP villages (98.7%) as compared to non-NGP village (98.4%) (Table 5.26).

Figure 5.7

Toilet Facility in Rayapurajupet Schools: AP

Source: Field Work.

Figure 5.8

In-House Septic Latrines in Gompa, AP, a Non-NGP Village

Source: Field Work.

Table 5.27

Frequency of Cleaning Toilets in NGP and Non-NGP Villages: AP

Frequency of Cleaning Toilets	*NGP*	*Non-NGP*	*Total*
Once or more in a day	38	27	65
Per cent	48.7	43.5	46.4
At least once in week	37	32	69
Per cent	47.4	51.6	49.3
At least once a fortnight	3	2	5
Per cent	3.8	3.2	3.6
At least once a month	0	1	1
Per cent	.0	1.6	.7
Total	78	62	140

Regarding the frequency of cleaning of toilets, at aggregate level, 49.3 per cent of the respondents clean their toilets at least once in a week, followed by 46.4 per cent of the respondents who clean toilets once or more in a day. Cleaning of toilets once or more in a day is a little higher in NGP villages (48.7%) as compared to non-NGP villages (43.5%). But, cleaning of toilets at least once in a week is higher in non-NGP villages (51.6%) as compared to NGP villages (47.4%). As far as cleaning of toilets at least once in a fortnight is concerned, there is not much of a difference between NGP and non-NGP villages (i.e. 3.8% and 3.2% respectively) (Table 5.27).

In this section, illnesses related to unsafe drinking water and unhygienic practices followed in the study villages resulting in health problems for children such as diarrhoea, dysentery and worm infection are discussed. Further, methods followed by the households for treating dehydration among children are also presented here.

With regards to seeking treatment for dehydration among children, 36.1 per cent of the respondents approach a doctor/s, followed by 32.4 per cent of the respondents who provide ORS (oral rehydration solution), and 27.5 per cent of the respondents provide ORS and also visit doctors. The percentage of the respondents visiting a doctor is more or less the same in respect of both NGP and non-NGP villages. A similar trend has been observed in both NGP and non-NGP villages with regard to providing ORS (Table 5.28).

Table 5.28

Treatment Seeking Behaviour of the Households with respect to Dehydration Among Children in NGP and Non-NGP Villages: AP

Measures Adopted	*NGP*	*Non-NGP*	*Total*
Provide ORS	40	39	79
Per cent	33.1	31.7	32.4
Provide only hot water	1	1	2
Per cent	.8	.8	.8
Visit a doctor/s	43	45	88
Per cent	35.5	36.6	36.1
Get the medicine directly from pharmacy	3	1	4
Per cent	2.5	.8	1.6
Approach Village Quacks	0	1	1
Per cent	.0	.8	.4
Provide ORS and provide only hot water	1	2	3
Per cent	.8	1.6	1.2
Provide ORS and visit a doctor/s	33	34	67
Per cent	27.3	27.6	27.5
Total	121	123	244

Table 5.29

Measures Taken by Households to Keep Cooked Food Safe in NGP and Non-NGP Villages: AP

Measures	*NGP*	*Non-NGP*	*Total*
Cover cooked food	121	123	244
Per cent	100.0	100.0	100.0
Total	121	123	244

In respect of both NGP and non-NGP areas, almost all the respondents always cover the cooked food with a plate (Table 5.29).

Regarding the frequency of taking bath by family members, at the aggregate level, in both NGP and non-NGP areas, a majority of the respondents (94.7%) have reported that they take bath once or more a day. In respect of NGP villages, 96.7 per cent of the respondents take bath daily as against 92.7 per cent of the respondents in non-NGP villages (Table 5.30).

Table 5.30

Household Level Personal Hygiene Practices: Taking Bath in NGP and Non-NGP Villages: AP

Frequency of Taking Bath	*NGP*	*Non-NGP*	*Total*
Once or more a day	117	114	231
Per cent	96.7	92.7	94.7
At least once in 2 days	4	9	13
Per cent	3.3	7.3	5.3
Total	121	123	244

VWSC and VHSC in the Study Villages: Awareness Level Regarding VWSC and VHSC among the Local Population

When enquired about the awareness of the existence of VWSC, in both areas together (NGP and non-NGP), only 9 per cent of the respondents have reported being aware of it. In NGP villages, about 15 per cent of the respondents are aware of VWSC, while in non-NGP villages, only 3.3 per cent of the respondents are aware of it. With regard to awareness of existence of the VHSC in their villages, 19.7 per cent of the respondents (at the aggregate level) have given a positive response. In respect of both NGP and non-NGP villages, the awareness level is the same in this respect (Table 5.31).

Table 5.31

Respondents Awareness Regarding the Existence of VWSC and VHSC in their Villages: AP

Existence of VWSC	*NGP*	*Non-NGP*	*Total*
Aware (VWSC)	18	4	22
Per cent (%)	14.9	3.3	9.0
Not aware (VWSC)	103	119	222
Per cent (%)	85.1	96.7	91.0
	Existence of VHSC		
Aware (VHSC)	24	24	48
Per cent (%)	19.8	19.5	19.7
Not aware (VHSC)	97	99	196
Per cent (%)	80.2	80.5	80.3
Total	121	123	244

An Analysis of Information Based on Child Health and Hygiene

As part of the study, an attempt was also made to collect information on various dimensions of child health and hygiene through administering an interview schedule to mothers of children aged 0-5 years. The purpose of the study/schedule was to find out health and hygiene practices followed by parents, especially mothers. The schedule covered various aspects such as practices of treating drinking water, awareness about diseases/ailments children may be vulnerable to due to unsafe drinking water, availability of toilet facility, especially for children, diseases/ailments affecting children due to practicing of open defecation or non-use of toilets, personal hygiene practices, improper bathing and washing of hands and material used for washing etc. The child schedules also contain questions related to the status of immunisation of children and maintenance of immunisation card separately for each child and illness episodes undergone by children during the last one year. The data on these aspects helped us have an idea not only about the health status of our future generations, but also the strengths and weaknesses of the policies meant for human development, especially improvements in respect of the health standards of children.

On being asked whether they treated drinking water before giving to their children, at the aggregate level, 51.2 per cent of the respondents have reported that they do treat water, whereas the remaining 48.8 per cent of the respondents do not treat water. In respect of NGP villages, the positive response is a tad higher (53.7%) as compared to non-NGP villages (48.8%). Among those who treat water before giving it to their children for drinking, at the aggregate level (out of 125), 88 per cent of the respondents have reported that they boil water followed by 8 per cent of the respondents who strain water using a cloth and 3.2 per cent of the respondents use water filters. In both NGP and non-NGP, water treatment measures are in the same proportion as at the overall level (Table 5.32).

Table 5.32

Distribution of Households Treating Water Before Consumption and Methods Making Water Safer for Children's Consumption: AP

Treating Water	*NGP*	*Non-NGP*	*Total*
Yes	65	60	125
Per cent	53.7	48.8	51.2
No	56	63	119
Per cent	46.3	51.2	8.8
Total	121	123	244
Practices to Make Water Safer to Drink			
Boil water	58	52	110
Per cent	89.2	86.7	88
Strain water, using a cloth	5	5	10
Per cent	7.7	8.3	8.0
Use water filters	2	2	4
Per cent	3.1	3.3	3.2
Boil and strain water, using a cloth	0	1	1
Per cent	.0	1.7	.8
Total	65	60	125

Table 5.33

Details of In-House Toilet Facility and Type of Toilets Used for Children in NGP and Non-NGP Villages: AP

Toilet Facility	*NGP*	*Non-NGP*	*Total*
Yes	78	63	141
Per cent	64.5	51.2	57.8
No	43	60	103
Per cent	35.5	48.8	42.2
Total	121	123	244
Type of Toilet Used for Children			
Toilet with in the household	76	62	138
Per cent	97.4	98.4	97.9
Others	2	1	3
Per cent	2.6	1.6	2.1
Total	78	63	141

The field data shows that at the aggregate level, 57.8 per cent of the respondents have access to in-house toilet facility and also gave positive response. In respect of NGP villages, 64.5 per cent of the respondents enjoy in-house toilet facility as against 51.2 per cent of the respondents in non-NGP villages. On being asked about the type of toilet facilities used (from among the respondents reporting having toilet facility) for their children, at the aggregate level, majority of them (97.9%) have reported the use of toilets. A similar trend has been observed, in both NGP and non-NGP villages, in this respect (Table 5.33)

Table 5.34

Distribution of Children Taking Help for Toilet Use in NGP and Non-NGP Villages: AP

Children Taking Help for Toilet Use	*NGP*	*Non-NGP*	*Total*
By his/her own	2	1	3
Per cent	1.7	.8	1.2
With the help of mother/father	119	122	241
Per cent	98.3	99.2	98.8
Total	121	123	244

Table 5.35

Awareness About Toilet Use and Children's Vulnerability to Possible Diseases in NGP and Non-NGP Villages: AP

Type of Diseases	*NGP*	*Non-NGP*	*Total*
Fever	5	18	23
Per cent	4.1	14.6	9.4
Fever & body pains	102	88	190
Per cent	84.3	71.5	77.9
Fever and chikungunya	1	0	1
Per cent	.8	.0	.4
Fever and body pains and chikungunya	13	17	30
Per cent	10.7	13.8	12.3
Total	121	123	244

With regard to the help of family members in the use of toilets by children at home, a majority of the respondents (NGP-98.3% and non-NGP-99.2%) have reported that their children take the help of parents in using toilets, while at the aggregate level it is 98.8 per cent (Table 5.34).

With regard to the possibility of exposure of children to disease in the event of them not using toilets facility, at the aggregate level, a majority of the respondents (77.9%) have reported that their children are prone to 'fever and body pain' followed by fever, body pains and chikungunya (12.3%). The percentage of respondents reporting fever and body pain is a tad higher in NGP villages (84.3%) as compared to 71.5 per cent that of the respondents in non-NGP villages (Table 5.34).

Figure 5.9

Health Awareness Meeting Conducted in ICDS Centre at Gompa Village: AP

Source: Field Work.

Table 5.36

Distribution of Respondents Reporting Frequency of Bath Given to Children in NGP and Non-NGP Villages: AP

Frequency of Giving Bath to Children	*NGP*	*Non-NGP*	*Total*
Once a day	11	22	33
Per cent	9.1	17.9	13.5
Twice a day	110	101	211
Per cent	90.9	82.1	86.5
Total	121	123	244

Regarding the frequency of bath given to the children, the field data shows that, at the aggregate level, a majority of the respondents (86.5%) give bath to their children twice a day, while the remaining 13.5 per cent per cent of the respondents once daily (Table 5.36).

Table 5.37

Distribution of Children with regard to Washing their Hands and Type of Wash Used before Taking Food in NGP and Non-NGP Villages: AP

Children Washing Their Hands	*NGP*	*Non-NGP*	*Total*
Yes	121	123	244
Per cent	100.0	100.0	100.0
No	0	0	0
Per cent	0.0	0.0	0.0
Type of Hand Wash Used for Children			
Water	53	59	112
Per cent	43.8	48.0	45.9
Soap	65	57	122
Per cent	53.7	46.3	50.0
Liquid soap	0	2	2
Per cent	.0	1.6	.8
Water and soap	3	5	8
Per cent	2.5	4.1	3.3
Total	121	123	244

Figure 5.10

Distributing Eggs to Children in ICDS Centre in Gompa Village: AP

Source: Field Work.

Almost all the respondents in both NGP and non-NGP areas have responded that their children always wash their hands, while according to more than half of the respondents (53.7% in NGP and 46.3% in non-NGP), their children wash their hands using soap. Close to another half of the respondents (45.9%) in both the areas have reported children washing their hands only with water (Table 5.37).

Table 5.38

Distribution of Households by Immunisation Services Received by Children in NGP and Non-NGP Villages: AP

Immunisation Status	*NGP*	*Non-NGP*	*Total*
Received	121	123	244
Per cent	100.0	100.0	100.0
Not received	0	0	0
Per cent	0.0	0.0	0.0
Total	121	123	244

According to almost all the respondents, their children have received immunisation (Table 5.38).

Illnesses Among Children in NGP Villages

In NGP areas, out of 164 children, 66 have been reported ill with waterborne diseases, 77 with other diseases which may be related to water and another 12 with other diseases not related to water.

Figure 5.11

Hand Wash Practices in ICDS in Gompa Village: AP

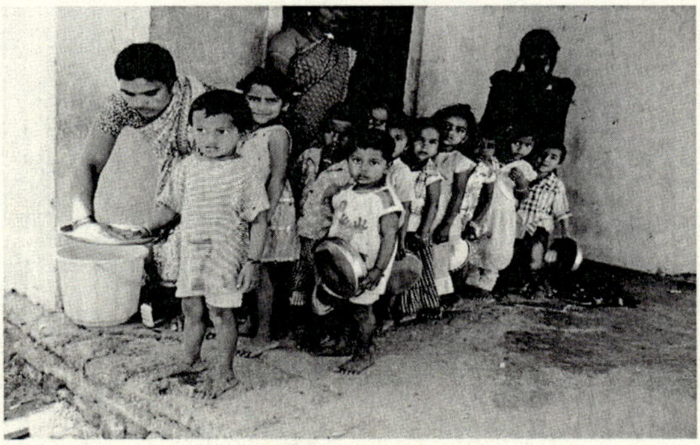

Source: Field Work.

Figure 5.12

Administering Vaccines to Children in Rajupeta Village: AP

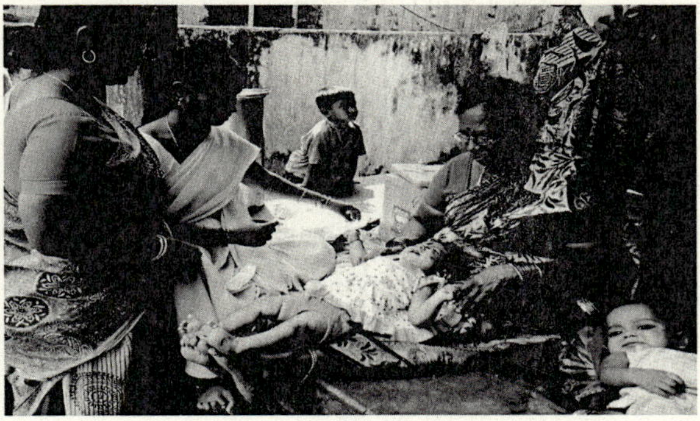

Source: Field Work.

Out of a total of 66 children reported suffering from waterborne diseases, 61 are dysentery cases, 9 malaria, and only 2 cases each of typhoid and diarrhoea and 1 case of jaundice. Out of 77 diseases-affected children (mostly related to water), 53 are reported suffering from fever and cold and 9 malaria cases (Tables 5.39 and 5.40).

Illnesses Among Children in Non-NGP Villages

In non-NGP villages, out of a total of 187 children, 95 children have been reported suffering from waterborne diseases; of them, 77 children suffering from dysentery, 3 children from diarrhoea, 6 children from malaria, while 2 each from cold and cough, jaundice and loose motion.

Out of 79 children afflicted with other diseases which may be related to water, 46 are reported to have suffered from fever and cold and cough, 28 from only fever and only 5 children from other diseases not related to water such as brain problem, snake bite etc.

Source of Treatment

In NGP villages, out of 61 dysentery cases, 38 patients are reported to have received treatment from private clinics, 8 each from SCs (sub-centres) and CHCs (community health centres);. 7 malaria patients, from private clinics, and out of 53 fever with cold and cough affected patients, 42 from private clinics and 6 from CHCs.

In non-NGP villages, a majority i.e., 58 dysentery-affected patients are reported to have taken treatment from private clinics; only 8 from PHCs (public health centres) and all the 3 diarrhoea affected patients from private clinics and out of 79, 35 fever and cold and cough affected and 24 fever affected patients from private clinics.

Treatment Duration and Expenditure

In NGP villages, a majority of 53 dysentery patients are reported to have undergone treatment for less than 5 days, with an average expenditure of ₹635.45, while in a few cases of malaria the length of treatment is reported to have lasted for more than 5 days, with an average treatment expenditure of ₹3,188.89. The treatment duration in the case of 44 cases (fever with cold) is reported to have lasted for less than 5 days with an average expenditure of ₹473.58.

In respect of non-NGP villages, the treatment days in the case of dysentery patients is reported to have lasted for less than 5 days with an average expenditure of ₹435.7 while the average expenditure in respect of diarrhoea cases amounting to ₹1,166.67, and the average expenditure in the case of jaundice patients to as high as ₹7,000, while the average expenditure for amebiasis to around ₹1,000. The average number of episodes for amebiasis is 1.

In both NGP and non-NGP areas, most of the children are reported to have been affected by dysentery followed by fever, cold and cough with children in both the cases taken to private clinics for treatment lasting for less than 5 days. It is clear from the results that there is clear advantage associated with water and sanitation programs in terms of reducing the incidences of water related diseases in NGP areas (Tables 5.39 and 3.40).

Table 5.39

Number of Children Reported Suffering from Illness in NGP Villages: AP

Disease Name		Af-fected	Average Episodes	Treatment Source					Treatment Duration (days)				Average Expendi-ture ₹
				SC	PHC	CHC	Private	SC & Private	< 5 Days	5 to 10	10 to 15	Above 15 Days	
Waterborne	Typhoid	2	1	0	2	0	0	0	0	1	0	1	1500
	Diarrhoea	2	1	0	0	0	1	1	2	0	0	0	1050
	Dysentery	61	1.31	8	2	8	38	5	53	8	0	0	635.45
	Jaundice	1	2	0	0	0	1	0	1	0	0	0	1500
Worm	0	0	0	0	0	0	0	0	0	0	0	0	0
Mosquito bite	Malaria	9	1.22	0	1	0	7	1	4	2	0	3	3188.89
Others (may be related)	Fever	13	1.69	0	0	1	11	1	11	2	0	0	561.54
	Viral fever	1	2	0	0	0	1	0	1	0	0	0	1000
	Fever and loose motion	3	1.67	0	0	0	3	0	3	0	0	0	1000
	Fever and fits	4	2	0	0	0	4	0	2	1	0	1	1625
	Fever and cold and cough	53	1.45	0	3	6	42	2	44	7	2	0	473.58
	Fever and cold and cough and infection	1	1	0	0	0	0	1	1	0	0	0	1000
	Fever and knee pain and stomach pain	2	2	0	0	0	2	0	1	1	0	0	1750
	Others		12										
	Total	164	1.45	8	8	15	110	11	123	22	2	5	2085.67

Table 5.40

Number of Children Reported Suffering from Illness in Non-NGP Villages: AP

	Disease Name	Affected	Average Episodes	Treatment Source					Treatment Duration (days)				Average Expenditure ₹
				SC	PHC	CHC	Private	SC & Private	< 5 Days	5 to 10	10 to 15	Above 15 Days	
Waterborne	Cold and cough	2	1	2	0	0	0	0.	2	0	0	0	1500
	Jaundice	2	1	0	0	0	2	0	0	0	0	2	7000
	Chikungunya	5	???	0	1	0	3	1	0	4	0	1	1000
	Typhoid	4	3.25	0	1	0	3	0	3	1	0	0	1175
	Dysentery	77	1.27	3	8	2	58	6	67	8	0	2	435.71
	Loose motions	2	4	0	0	0	2	0	1	1	0	0	1750
	Diarrhoea	3	1	0	0	0	3	0	2	0	1	0	1166.67
Worm	Amebiasis	1	1	0	0	0	1	0	0	0	0	1	1000
	Nulipurugulu	1	2	1	0	0	0	0	1	0	0	0	1000
	Malaria	6	2	0	1	0	5	0	1	4	0	1	733.33

Other (related)	Fever	28	1.25	1	2	0	24	1	25	3	0	0	407.69
	Fever and headache	1	1	0	0	0	1	0	1	0	0	0	200
	Fever and cold and cough	46	1.43	2	3	3	35	3	26	19	0	1	773.86
	Fever and cold and cough and headache	1	1	0	0	0	0	1	1	0	0	0	400
	Fever, cold and cough, knee pain and stomach pain and vomitings	1	1	0	0	0	1	0	0	1	0	0	600
	Fever and knee pain and stomach pain	2	3	0	0	0	2	0	1	1	0	0	850
	Others	5	1	0	0	0	1	0	0	0	0	1	20000
	Grand total	187	1.42	9	16	5	145	12	132	88	3	40	2425.29

A Profile of Diseases Among Adults

In NGP areas, 182 out of 652 adults reported being affected by some waterborne diseases, 42 are stated to have suffered from other diseases which may be related to water, while the remaining 56 from other diseases such as TB, BP, eye problem, kidney problem etc., which are not at all related to water and sanitation. Out of 84 adults reported with waterborne diseases in NGP villages, 47 come under dysentery, 16 under malaria, 10 under typhoid, one case under loose motion and digestion problem, 5 under diarrhoea, and 2 cases each under jaundice and amebiasis. Out of 42 adults reported suffering from other diseases which may be related to water, 25 come under fever.

In non-NGP areas, out of a total of 224 adults suffering from some kind of diseases, 104 have been reported afflicted with waterborne diseases, 34 with other diseases which may be related to water and 65 with other diseases such as cancer, heart problem, HIV, kidney, etc., which are not at all related to water and sanitation. Out of the 104 adults affected by waterborne diseases, 56 are reported having dysentery, 19 case of typhoid, while only one case each of loose motion and jaundice and three cases of diarrhoea have been reported. Out of the 34 adults with other diseases which may be related to water, all have had fever—23 only fever and the remaining 11 having fever with other diseases like kidney problem (Table 5.42).

Source of Treatment

In NGP villages, out of 47 dysentery patients, 11 are reported to have approached SC for treatment and 26 private clinics; out of 16 malaria cases, 9 are reported to have received treatment from private clinics; out of 10 typhoid patients, 6 from private clinics; out of 25 cases of fever, 16 from private clinics.

In non-NGP areas, out of 56 dysentery-affected adults, 38 went to private clinic while 9 went to PHC, 5 to SC, 1 to CHC and 3 to both SC and private. Out of 19 adults down with typhoid, 12 are reported to have received treatment from private clinics, 5 from PHCs, and 2 from both PHCs and private clinics (Tables 5.41 and 5.42).

Table 5.41

Illness Episodes Encountered, Treatment Sought and Expenditures Incurred by Households in NGP Areas: AP

	Disease Name	Affected	Average Episodes	Treatment Source					Treatment Duration (days)				Average Expenditure (₹)
				SC	PHC	CHC	DH	Private	SC & Private	< 5 Days	5 to 10	Above 15 Days	
Waterborne	Typhoid	10	1.7	0	2	0	1	6	1	3	4	3	2800
	Dysentery	47	1.06	11	2	6	0	26	2	44	3	0	328.38
	Diarrhoea	5	1.4	0	0	0	0	4	1	5	0	0	1110
	Motions	1	3	0	0	0	0	1	0	1	0	0	2000
	Jaundice	2	2	0	0	0	0	2	0	0	0	2	7500
	Digestion problem	1	1	0	0	0	0	0	1	0	0	1	1500
Worm infections	Amebiasis	2	1	0	0	0	0	0	2	1	1	0	200
Mosquito bite	Malaria	16	1.56	0	4	0	0	9	3	3	10	3	2900
Others but related to water	Chikungunya	1	1	0	0	0	0	1	0	1	0	0	2000
	Fever and leg pain	2	2	0	0	0	0	1	1	0	1	1	11000
	Knee pain and stomach pain	8	1.5	0	0	0	0	3	5	2	1	5	18400
	Fever	25	1.44	3	4	1	0	16	1	21	4	0	422.73
	Viral fever	3	1.67	0	0	0	0	3	0	0	3	0	1166.67
	Fever and cold and cough	2	1	0	0	0	0	0	2	0	0	2	15750
	Fits	1	2	0	0	0	0	1	0	0	0	1	25000
	Others	56											
	Grand total	182	1.46	14	12	7	1	73	19	81	27	18	3553.24

Table 5.42

Illness Episodes Encountered, Treatment Sought and Expenditures Incurred by Households in Non-NGP Areas: AP

	Disease Name	*Affected*	*Average Episodes*	*Treatment Source*					*Treatment Duration (days)*					*Average Expenditure ₹*
				SC	*PHC*	CHC	DH	Private	SC & Private	< 5 Days	5 to 10	10-15	Above 15 Days	
Waterborne	Cold and cough	10	1.80	0	0	4	2	4	0	0	5	2	3	2320.00
	Typhoid	19	2.05	0	5	0	0	12	2	6	7	1	5	1805.26
	Motions	1	2	0	0	0	0	1	0	0	1	0	0	1000
	Dysentery	56	1.07	5	9	1	0	38	3	53	6	0	0	318.33
	Chikungunya	14	1.29	0	1	0	0	13	0	1	5	2	6	1984.62
	Diarrhoea	3	2	1	2	0	0	0	0	2	0	1	0	1025
	Jaundice	1	3	0	1	0	0	0	0	0	0	0	1	6000
Worm	Amebiasis	2	1	0	1	0	0	0	1	2	0	0	0	225
Mosquito bite	Malaria	19	1.63	0	2	1	0	13	3	6	6	4	3	1565.79
Others-may be related	Fever	23	1.22	1	6	0	0	14	2	17	6	0	0	481.58
	Fever and kidney problem	2	1	0	0	0	0	2	0	2	0	0	0	400

Fever, and knee pain and stomach pain	1	4	0	0	0	0	1	0	1	0	0	0	700
Fever, knee pain, stomach pain and head ache	1	1	0	0	0	0	1	0	1	0	0	0	700
Fever and leg pain	2	1.5	0	0	0	0	1	1	0	2	0	0	700
Fever, cold and cough, stomach pain and knee pain	2	1.5	0	0	0	0	2	0	1	1	0	0	325
Fever and body pains	2	1.5	0	0	0	0	2	0	0	2	0	0	1250
Fits	1	4	0	0	0	0	1	0	0	1	0	0	20000
Others	65												
Grand total	224	1.49	7	27	6	2	105	12	92	42	10	18	3054.67

Details of Treatment: Duration and Expenditure

In NGP areas, the treatment of 44 dysentery cases is reported to have lasted for less than 5 days, while in only 3 cases of malaria for less than 5 days with an average expenditure of ₹328.38 and ₹2,900 respectively. The average expenditure incurred on typhoid amounts to ₹2,800 and in most of the fever cases, the treatment is reported to have lasted for less than 5 days, with an average expenditure of ₹422.73.

In non-NGP areas, the number of treatment days for dysentery cases is reported to have lasted for less than 5 days with an average expenditure of ₹318.33. The average expenditure incurred on typhoid patients amounts to ₹1,805.26. In the case of jaundice, the child experienced 3 episodes with the health expenditure being as high as ₹6,000. Out of 23 reported with fever, 14 are reported to have treatment from private clinics and 6 from PHCs. In most of the fever cases, the number of days for treatment has been reported less than 5 with the average expenditure being ₹481.58. The treatment days for malaria patients have been reported lasting between 5-10 days.

In both NGP and non-NGP areas, a maximum number of adults are reported to have suffered from dysentery followed by malaria and typhoid, with the treatment in all the cases lasting mostly for less than 5 days. Adults suffering from any kind of diseases in both NGP and non-NGP areas, mostly visit private clinics. Compared to NGP areas, non-NGP areas have reported more number of water related illnesses. It shows the positive impact of water and sanitation programmes on the health of individuals.

Summary of Discussions Based on Focus Group Discussions (FGDs) in Both NGP and Non-NGP Villages in AP

Main Source of Water Supply

- In all NGP villages, the main source of water supply is tap water supplied by the *panchayat* concerned. All the households have access to water supply in the villages.
- In four non-NGP villages (Turlapudi, Gotivad Agraharam, Gompa, and Ramachandrapalem), tap water is the main

source of water. In Kothali and Damunapalle villages, the main source of water supply is from panchayat water taps.

Adequacy of Water

- In all the NGP villages, there is a sufficient water supply. Only during summer, water supply gets disrupted due to an erratic power supply.
- In all non-NGP villages, water supply is sufficient during all seasons. However, during summer, villagers face water problem due to frequent power cuts. During rainy season, water turns to red colour.

Quality of Water

- In all NGP villages, water is clean and good and once in a month, water tanks are cleaned with bleaching powder.
- In all non-NGP villages, water tanks are cleaned with bleaching powder.

Water Costs

- In Rayapurajupeta, L. Singavaram, Lakkavaram, Kavagunta and Rajupeta NGP villages, people do not pay for water consumption/supply, while in V.J. Puram village, people pay a nominal amount for water used.
- In all non-NGP villages, people do not pay money for water supply.

Health Problems

- In Rayapurajupeta NGP village, most of the people suffer from fever during rainy season and in L. Singavaram village, they suffer from fever and cough. In V.J. Puram village, people tend to get fever, cold and cough.
- In Damunapalle, people suffer from diarrhoea, fever and other diseases. In Gotivad Agraharam village, a majority of people suffer from fever. In Gompa village, people have reported malaria, typhoid, stomach pain, motions and 'nulipurugulu' (worms).

Sanitation

- In Kavagunta and Rayapurajupeta NGP villages, some houses have toilet facility. In Lakkavaram (80%) and in Rajupeta (95%) villages, a maximum number of households have access to toilet facility. In V.J. Puram village, half of the houses have toilet facility, whereas in L. Singavaram village, only 20 houses have toilet facility.
- In Turlapudi, Damunapalle and Gotivad Agraharam Non-NGP villages, most of the households do not have access to toilet facility, while in Gompa village, 100 houses have toilets. In Kothali and Ramachandrapalem villages, 20-30 per cent of the houses have toilet facility.

Open Defecation

- In Rayapurajupeta, L. Singavaram, V.J. Puram and Lakavaram villages, 30-50 per cent of the people, in Kavagunta village, 95 % of the people, and in Rajupeta village (all NGP villages), 5% of the people practise open defecation. People tend to face problems using toilets in Rayapurajupeta village, as tanks over flow during rainy season, while in Kavagunta village, there are problems in using toilets during summer season due to an inadequate water supply.
- In all the non-NGP villages, 80 to 95 per cent of the people practice open defecation. Those having toilets in villages like Turalapudi and Gompa, face problems in using toilets due to an inadequate water supply in the villages during summer. In Gompa village, as they do not have toilets within households, people face problems during rainy season. They also face problems at night, as they need walk long distances.

Problems Due to Open Defecation

- In L. Singavaram NGP village, according to people, fever, diarrhoea and communicable diseases are common; Malaria, dengue, cholera, chikungunya are common among children aged 0-5 years in V.J. Puram village; in Lakkavaram village, children tend to face health problems like vomiting, cholera, diarrhoea and typhoid, mainly because of the prevalence of open defecation; In Kavagunta village, children are prone to face cold, cough, fever, malaria, dengue, typhoid, and itching; in Rajupeta village, according to people, children are suffering from communicable diseases, respiratory problems, vomiting and diarrhoea.

- In Turlapudi and Ramachandrapalem, non-NGP villages, children suffer from cold, cough, fever, malaria, dengue, typhoid and itching; in Gompa village, children suffer from fever, skin infections etc. In Kothali village, health problems faced by children include cholera, communicable diseases, itching and rashes.

Functioning of Committees—VWSC

- In Rayapurajupeta and Rajupeta NGP villages, people are not aware of VWSC. The committees in L. Singavaram and V.J. Puram villages supervise activities like cleaning of water tanks and garbage disposal. The VWSC in Lakkavaram village takes up works like mixing bleaching powder with water, maintaining taps and cleaning streets. In Kavagunta village, pouring of bleaching powder on the sides of streets and water tanks is taken up on a regular basis.
- In Turlapudi, Gotivad Agraharam, and Ramachandrapalem non-NGP villages, people are not aware of these committees. In Turlapudi village, committees have taken up works like pouring bleaching powder on the sides of streets and water tank cleaning. The committee in Gotivad Agraharam village has taken up works like cleaning of water tanks with bleaching powder and repairing of water taps; streets are cleaned on time and regular spraying of bleaching powder in the village is done.

VHSC

- In all NGP villages, VHSC are working. In L. Singavaram village, the committee creates awareness among people regarding health issues. In Rajupeta village, the VHSC creates awareness regarding diseases besides educating people about personal hygiene. In L. Singavaram village, Accredited Social Welfare Activist (ASHA) worker educates people regarding personal hygiene and measures to be taken to prevent diseases. In V.J. Puram village, ASHA advises every individual household to have toilet facility at home. In Lakkavaram village, ASHA worker creates awareness among people regarding hygiene and measures to be taken to prevent diseases. In Kavagunta village, ASHA worker creates awareness regarding personal hygiene and measures to be taken to prevent diseases.

- In all non-NGP villages, the health committees are functioning. In Gotivad Agraharam and Gompa villages, VHSCs are very useful in creating awareness among people regarding personal hygiene.

Suggestions to Improve the Performance of VWSC and VHSC

- In Rayapurajupeta NGP village, the committees have to be formed. In V.J. Puram village, roads should be cleaned. Everyone in the village should use toilets. In Lakkavaram and Kavagunta villages, training needs to be given to the committee members. In Rajupta village, spraying of bleaching powder should be taken up besides creating awareness among people regarding hygiene.
- In Turlapudi, Gotivad Agraharam and Kothali non-NGP villages, training and regular meetings should be conducted and the committees should educate people about hygiene conditions. In Damunapalle village, committees should keep the village clean besides conducting awareness camps.

Summary

As discussed in the introduction chapter, good health outcomes depend on various interventions. When the health programmes are formulated and implemented effectively, morbidity levels in the villages may come down and diarrhoea episodes may get reduced and there may be better nutrition absorption among children etc. Social determinants considered in this study, included the availability of adequate quantity and quality water, and sanitation facilities (in this case, lavatories and their proper use by households), allied household level hygiene practices conducive to promoting health, personal hygiene such as defecation habits. This kind of enabling environment is made possible indirectly through a catalytic process involving interactions among various institutions, processes and different programmes at the local level. The study is basically explanatory and analytical in nature.

Here an attempt is made to analyse and understand health conditions of individuals and households as these are linked to access to water and sanitation facilities as well as their health and hygiene awareness, practices at the individual and household levels.

Water: In AP, 6 NGP and 6 non-NGP villages were selected for the study. From these 12 villages, data was collected from 244 households. In this study, an attempt is made to see better health conditions due to the implementation of water and sanitation programmes. A comparison of NGP and non-NGP villages indicates that pure water availability is more in NGP villages. The availability of an adequate quantity of water to the people is also one of the parameters to understand the determinants of health and hygiene conditions of the sample households. At the aggregate level, a majority of the persons live in *pucca* houses followed by semi-*pucca* houses (10.2%). In NGP and non-NGP also, the proportions remain nearly the same. Tap water is the main source of water in both NGP and non-NGP villages. As regards using the occasional source of water, out of 244 respondents, a majority of the respondents use water for domestic purposes only.

In NGP villages, out of six schools, two schools get drinking water from taps, and one each from tanks, cans and public taps. Out of eight schools in non-NGP villages, three schools get drinking water from taps and another three from tanks.

At the aggregate level, a majority of the respondents have reported that the quantity of water is fully sufficient. In respect of NGP villages, the 'fully sufficient' response is higher (88.4%) than in non-NGP villages (82.1%). At the aggregate level, most of the respondents have reported that water supply has been sufficient during the past 1 year. However, in NGP villages, such a positive response is a little higher (86.8%) as compared to non-NGP villages (82.9%). In the sample areas of AP, a majority of the households use earthen pots besides plastic/steel vessels for water storage. In summer, water problem arises mainly due to an erratic power supply.

At the aggregate level, out of 244 respondents, 98 per cent take water from containers with glass tumblers/utensils using hand. The respondents who 'sometimes' treat drinking water are slightly more in number in non-NGP (46.3%) relative to NGP villages (43.8%). Out of 125 respondents who take measures to make water safer for drinking, 91.2 per cent boil water, followed by 7.2 per cent who strain water using a cloth. In NGP villages, the percentage of respondents boiling water is higher (93.4%) as compared to those in non-NGP villages (89.1%). In NGP villages, the percentage of respondents giving boiled water to their children is a little higher (89.2%) than those in non-NGP villages (86.7%). In both NGP and non-NGP areas, almost all the respondents cover cooked food with a plate.

Sanitation: In NGP areas, two thirds of the respondents use septic latrines and one third of them practice open defecation. In non-NGP areas, half of the respondent's practise open defecation, while nearly another half of them uses septic latrines. In both NGP and non-NGP villages, respondents use toilets for maintaining better health. The percentage of respondents washing hands (after defecation) with soap is slightly higher in respect of NGP villages (37.2%) as compared to non-NGP villages (30.9%). Washing hands with only plain water is more visible in non-NGP villages (60.2%) as compared NGP villages (54.5%).

As regards the availability of latrine facilities in schools, all the six schools in NGP have septic latrines, whereas only seven out of eight schools in non-NGP villages have septic latrines. All the six schools in NGP villages have separate toilets for boys, girls and teachers, whereas in non-NGP villages, in five out of eight schools there is such a facility.

Figure 5.13

Proverbs Written on Walls about the Importance of Toilet Use in Hazipalli: AP

Source: Field Work.

Figure 5.14

Every Household has Toilet Facility in Hazipalli (NGP) in AP

Source: Field Work.

For cleaning toilets in schools, in NGP villages, a majority of the schools (four out of six), hire cleaners and sweepers. In non-NGP villages, half of the schools (four out of eight schools), hire sweepers and cleaners for cleaning toilets. In NGP, in a majority of schools toilets are cleaned fortnightly (four out of six schools), while in non-NGP villages, in three out of eight schools toilets are cleaned fortnightly and followed by daily in one school and weekly in one school.

Water supply in schools: Water supply is sufficient in five out of six schools in NGP, while it is so only in half of the villages (four out of eight schools) in non-NGP villages.

Sanitation in schools: About the type of toilets facilities used (from among the respondents reporting the availability of toilet facility) for their children, at the aggregate level, a majority of them (97.9%) have reported the use of toilets within the household premises. In respect of both NGP and non-NGP villages, the situation is more or less similar in this respect.

About disposing of children's stools, a majority of the respondents throw on the street (73.8%), followed by 'thrown in latrine' (19.3%) and 'left there' (7%). In NGP villages, the percentage of the respondents throwing stools in the latrine is higher (24%) as compared to those in non-NGP villages (14.6%). A majority of the respondents in both the areas have reported that their children get the help of parents in using toilets. As regards washing hands, nearly two thirds of the respondents have stated that children wash their hands both before and after taking food (64% in non-NGP and 62.8% in NGP villages). According to more than half of the respondents (53.7% in NGP and 46.3% in non-NGP), their children wash their hands using soap. With regard to seeking treatment for dehydration among children, 36.1 per cent of the respondents visit a doctor and another 32.4 per cent of them administer ORS. In NGP villages, relatively more number of respondents takes bath daily (96.7% in NGP and 92.7% in non-NGP). A majority of the respondents (86.5%) in both the areas give bath to their children twice a day.

Institutions: Here an attempt is made to understand the role of institutions like VWSC, VHSC, SHGs (self help groups), CBOs (community based organisations) etc. and the processes adopted for achieving ODF (open defecation free) villages and the provision of clean drinking water to all. An attempt is also made to understand the role of institutions and processes that tend to hinder the achieving of ODF status of villages and supply of drinking water.

In both the types of villages, there are VWSCs. As regards the frequency of meetings, in 4 NGP villages, quarterly meetings are held in 2 villages and monthly/half yearly meetings are held in other 2 villages. In all, 4 villages in NGP, maintain records of the meetings and only one village in non-NGP keeps records of those meetings. As regards the frequency of VHSC meetings, mostly monthly meetings are held in both NGP and non-NGP areas. Records of such meetings are maintained in all the villages of NGP and non-NGP areas.

Illnesses among adults: In both NGP and non-NGP areas, a maximum number of adults have been reported suffering from dysentery followed by malaria and typhoid. In all the cases, treatment days are mostly less than 5. Adults suffering from any kind of diseases in both NGP and non-NGP areas, mostly visit private clinics for treatment. Compared to NGP areas, non-NGP areas have reported more number of water related illnesses. It shows the positive impact of water and sanitation programmes and working of village level institutions like VWSC and VHSC on the health of individuals.

Illnesses among children: In both NGP and non-NGP areas, most of the children have been reported down with dysentery followed by malaria. In both the cases, children are reported to have been taken to private clinics for treatment, lasting for less than five days. As in the case of health of adults, it is clear from the results that there is a clear advantage of water and sanitation programme in terms of reducing the incidences of water related diseases in NGP areas.

6

A Comparison of Findings across the States of Madhya Pradesh, Andhra Pradesh and Odisha

As discussed in the introduction chapter, good health outcomes, in a broad sense, depend on different interventions-direct health interventions and other institutional interventions like VWSC (village water and sanitation committee), VHSC (village health and sanitation committee), SHGs (self help group) and other CBOs (comminity based organisation) at different levels. When the health programmes are formulated and implemented effectively, morbidity levels in the villages may come down and may reduce episodes of diarrhoea and other related diseases and may lead to better nutrition absorption among children etc. Further, the health outcomes largely depend on how the interactions take place among various institutions, such as VWSC, VHSC, SHGs and other CBOs, the processes and different programmes. Having analysed the state-wise (Madhya Pradesh [MP], Andhra Pradesh [AP] and Odisha) situation in the previous chapters, the present chapter attempts to draw a comparative perspective of all the three states, especially with respect to water and sanitation facilities influencing the health outcomes in the three states. The essential focus in this chapter will be on access issues regarding water and sanitation and its utilisation pattern as well as the reasons for non-utilisation; the presence and absence of the local level water and sanitation committees, the processes adopted for their working and their implications for the health outcomes of communities.

Main Sources of Water across the Three States

Regarding the main sources of water, comparatively both Nirmal Gram Puraskar (NGP) and non-NGP villages in AP and NGP villages in Odisha have a better access to safe drinking water. The sources of water, to a large extent, decide to what extent people have a ready access to safe and potable drinking water. Across the three states studied, in MP, in a majority of the villages coming under both NGP and non-NGP villages, the main source of water is hand pumps, whereas in Odisha, a majority of the villages have piped water connection to the public taps and tap connection to houses in NGP villages. In non-NGP areas, the main source of water is tube well/bore hole. In AP, in a majority of the villages, tap water is the main source of water in both NGP and non-NGP villages. On the whole, it shows that in AP (both NGP and non-NGP) and NGP villages in Odisha, access to safe drinking water is better relative to villages in MP (Figure 6.1).

Figure 6.1

Main Sources of Water in Three States

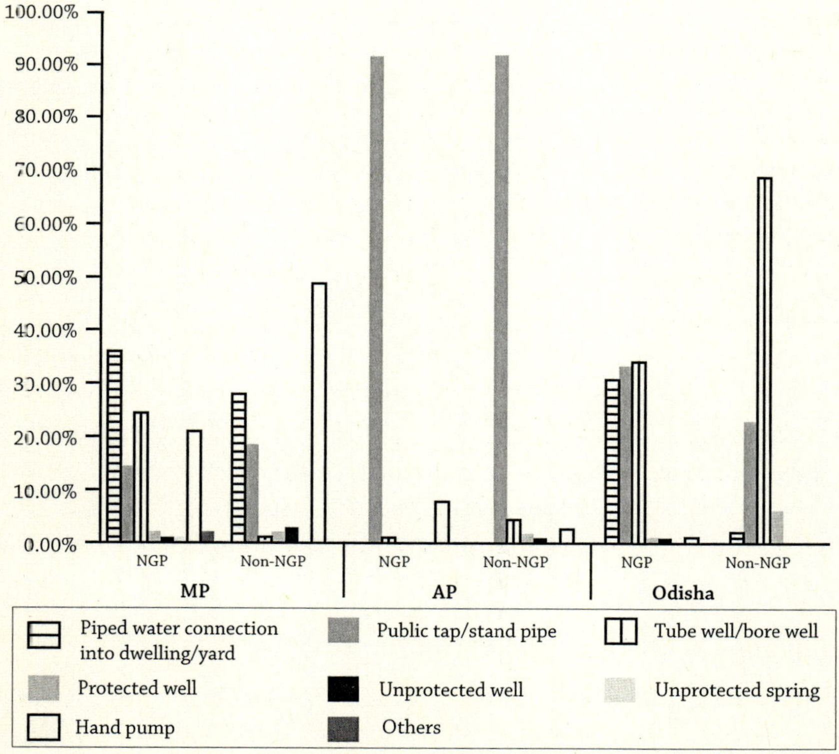

Quantity of Water Supply

The quantity of water supply is sufficient for many households in both NGP and non-NGP villages of MP and AP, whereas its availability is a problem in non-NGP villages of Odisha.

Figure 6.2

Perceptions of the Respondents Regarding the Quantity of Water Available across the Three States

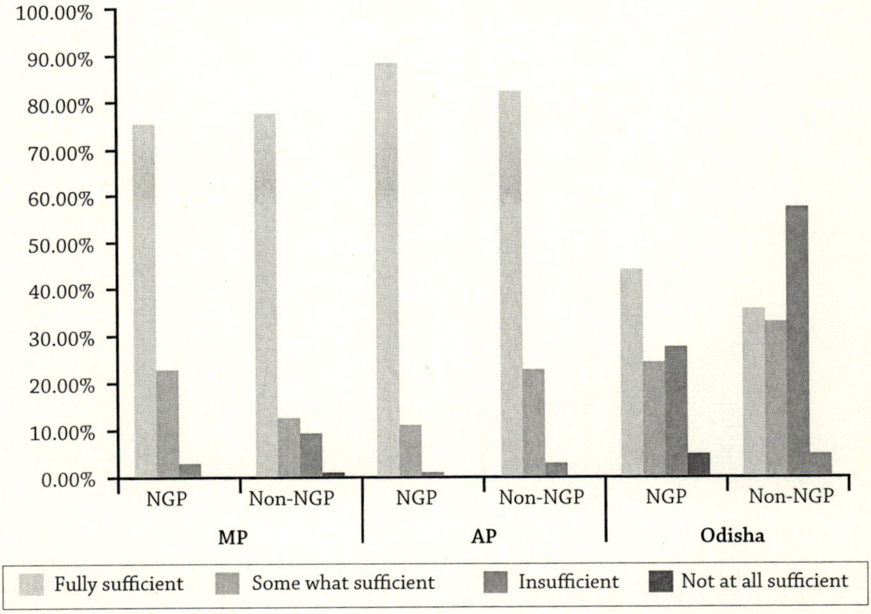

The availability of water is sufficient in both NGP (75%) and non-NGP areas (77.3%) in MP. In Odisha state, a majority of the respondents in NGP villages (43.8%) have reported that the quantity of water is fully sufficient, whereas in non-NGP villages, it is only 5 per cent. A large numbers of respondents in NGP villages have access to an adequate water as compared to those in non-NGP villages. Only 24.6 per cent in NGP and 32.8 per cent of the respondents in non-NGP areas have stated that water supply is somewhat sufficient.

In AP, a majority of the respondents (85.2%) have reported that the quantity of water is fully sufficient. In respect of NGP villages, the 'fully sufficient' response is higher (88.4%) than in non-NGP

(82.1%) areas. At the aggregate level, most of the respondents have reported that water supply has been sufficient during the past one year. However, in NGP villages, this response is a little higher (86.8%) relative to non-NGP villages (82.9%) (Figure 6.2).

Quality of Water

As far as quality of water[1] is concerned, a majority of the households in AP (both NGP and non-NGP villages) have reported that water is pure as compared to those in Odisha (both NGP and non-NGP villages),whereas in MP, households have reported

Figure 6.3

Quality of Water across the Three States

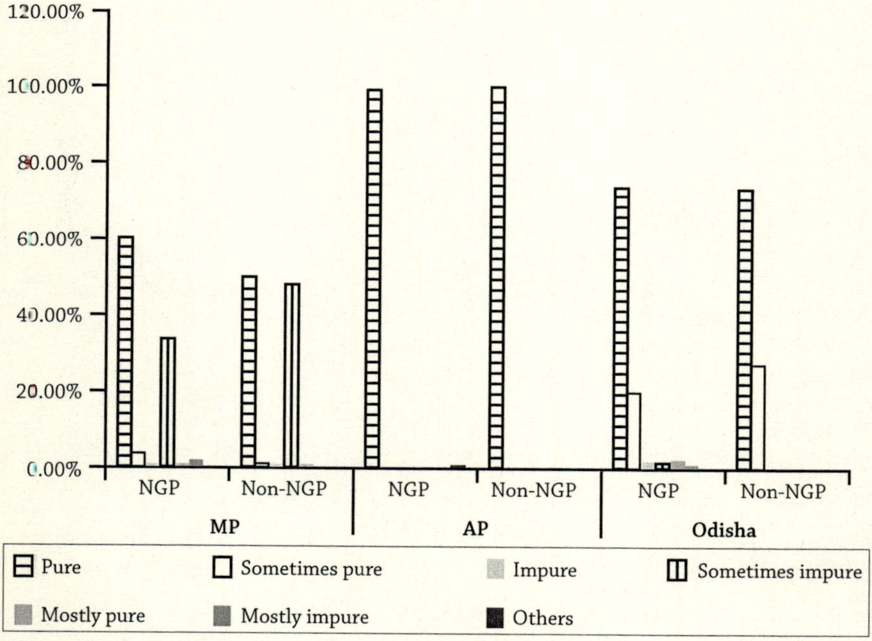

problems regarding the quality of water. Besides the quantity of water available, the quality of water is also an important issue, considering that consuming impure water can cause a series of health problems. There are a number of water quality issues in MP like 'hard' water

1. Purely based on physical appearance and the perception of respondents.

and water being yellowish in colour, while in Odisha three fourths of the respondents in both NGP and non-NGP have stated that water is pure. Some respondents (2.08%) have reported that water is impure, especially during the rainy season. In AP, almost all the respondents in both NGP and non-NGP areas have reported that water is pure (Figure 6.3).

Water Related Hygiene Practices

Compared to Odisha and MP, almost all the households in AP have reported cleaning of water containers daily. Keeping drinking water containers clean is quite important as using unclean containers for storing water can cause a number of diseases among people. An

Figure 6.4

Frequency of Cleaning Water Containers

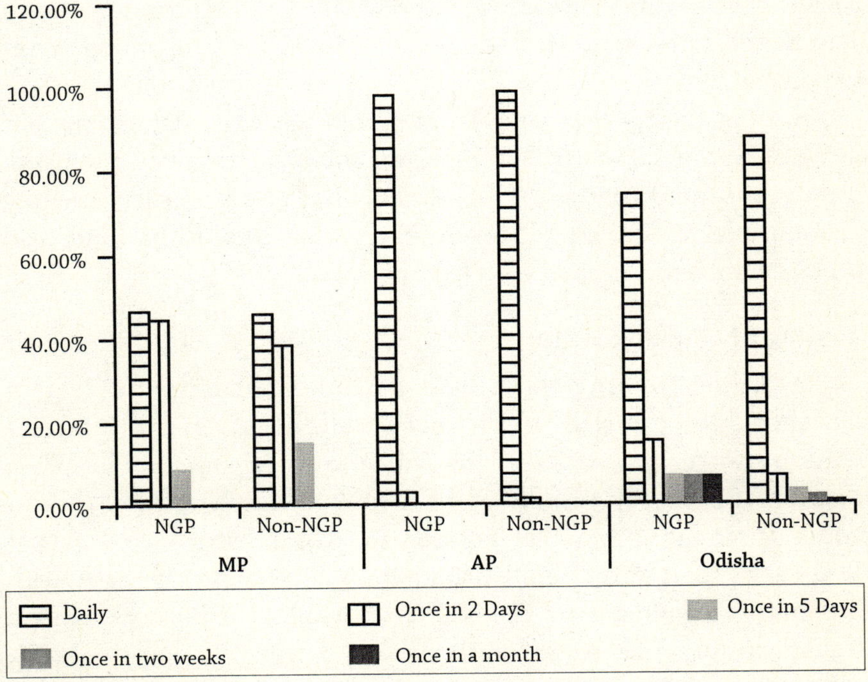

analysis across the 3 states reveals that, in MP, 45.4 per cent of the respondents in non-NGP villages clean drinking water containers daily. In NGP villages, 45.8 per cent of the respondent's clean containers daily, whereas in Odisha, those who clean containers every day, are more in number in non-NGP villages (87.4%), as compared to NGP villages (74.4%). In AP, a majority of respondents clean water containers daily. In non-NGP villages, 98.4 per cent of the respondents clean water containers daily as against 97.5 per cent of the respondents in NGP villages (Figure 6.4).

Practices of Hygiene Maintenance

Hygiene maintenance practices relating to drinking water are relatively better in NGP areas of MP. For example, the respondents who use glass tumblers attached with long handles (to draw water from drinking water container) to avoid contamination are more in number in NGP areas.

In Odisha state, a little over one fifth in NGP areas and less than 10 per cent of the respondents use glass tumblers attached with long handles and taps attached to containers for consuming water respectively.

In AP, at the aggregate level, out of 244 respondents, 98 per cent take water from containers with glass tumblers/utensils, using hand. It is to be noted that in AP—both in rural and urban areas—the common practice is using glass tumblers/utensils with hand instead of glass tumblers attached with long handles.

Treating of Drinking Water

Regarding treating water before drinking, a large number of households in both NGP and non-NGP villages of AP and MP treat drinking water as compared to households in Odisha. In MP, a majority (92) of the respondents in NGP areas always treat water before drinking. In non-NGP villages, 82 respondents have told that they always treat water before drinking. In Odisha, nearly three forth of them have replied that they boil water, while a little over one fifth of them strain water using a cloth.

In AP state, the percentage of respondents who 'sometimes' treat drinking water is a little higher in non-NGP villages (46.3%) than those in NGP villages (43.8%). Out of 125 respondents, who take measures to make water safer before drinking, 91.2 per cent boil water followed by 7.2 per cent of the respondents who strain water, using a cloth. In NGP villages, 'boil water' response is higher (93.4%) as compared to non-NGP villages (89.1%).

Type of Housing Facility

Regarding housing, a majority of the households in AP (both NGP and non-NGP villages) live in *pucca* houses relative to their counterparts in the other two states (MP and Odisha), whereas, most of the respondents in both the areas of MP live in huts or *kutcha* houses. It is generally perceived that people living in semi-*pucca* or *pucca* houses are relatively better off and that they can afford better sanitation facilities.

Figure 6.5

Type of Housing Facility

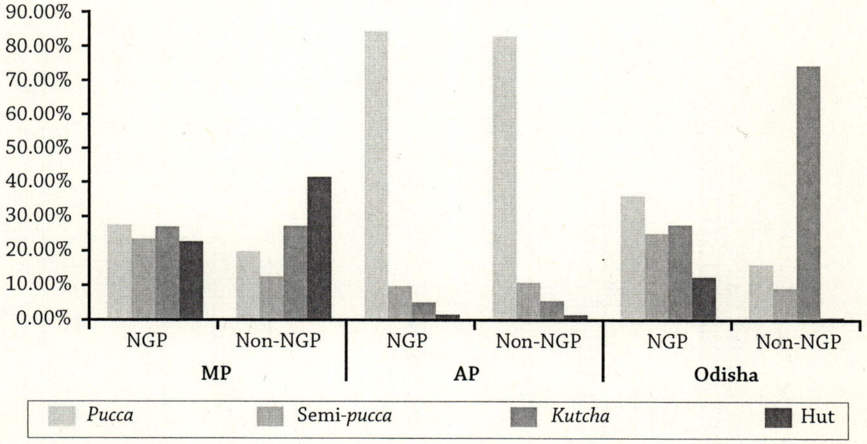

In Odisha state, many respondents reside in *kutcha* houses and huts, while most of the respondents belonging to non-NGP villages, and live mostly in *kutcha* houses and huts, indicating their poor economic conditions. In AP, a majority of the persons have *pucca* houses. In NGP and non-NGP villages also, the proportions remain nearly the same (Figure 6.5).

Sanitation Facility

Having toilet facilities within the household premises, go a long way in maintaining a proper health of the household members. Overall, the availability of toilet facility is better in AP and MP, when compared to Odisha state. The state-wise breakup of data regarding the availability of toilets facility and its usage is given as follows.

In MP, out of 239, 140 households in both NGP and non-NGP areas (96 from NGP and 44 from non-NGP) reported having in-house toilets. Whereas in Odisha, both in NGP and non-NGP villages, out of 240 households, 164 (68.3%) households practice open defecation and 76 (31.7%) have access to toilet facility.

Figure 6.6

Sanitation Facility

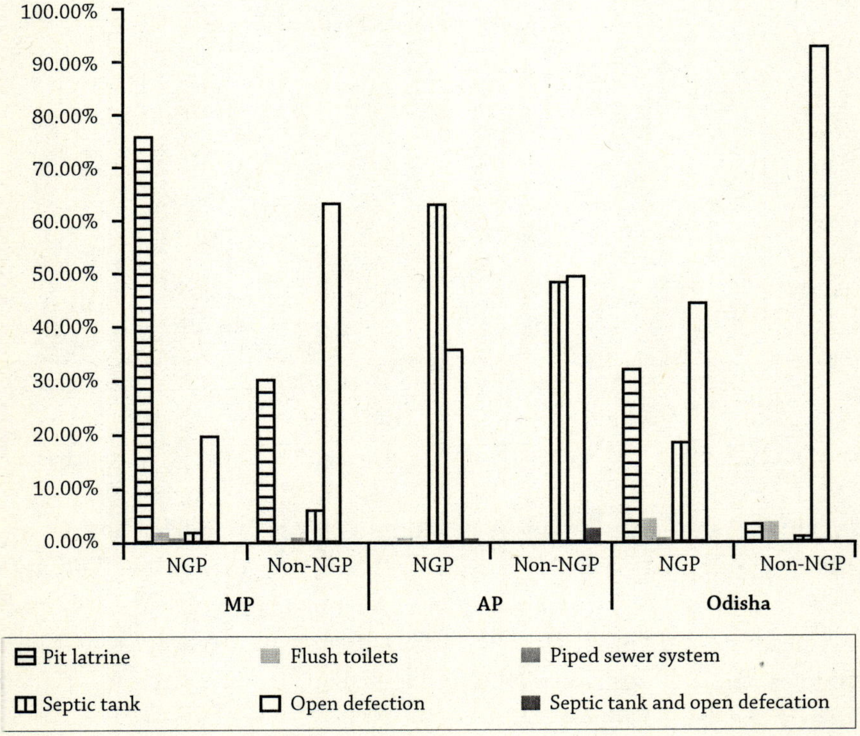

In NGP areas of AP, two third of the households use septic latrines and one-third practice open defecation. In non-NGP areas, half of the respondents practice open defecation and nearly another half of them use septic latrines. More than half (57.4%) of the respondents have access to in-house toilet facility, while the rest of them (42.6%) use open spaces for open defecation (Figure 6. 6).

Frequency of Cleaning Toilets

Regarding cleaning of toilets, a majority of the households (with toilets) in AP and MP, clean their toilets regularly as compared to their counterparts in Odisha. Improper maintenance of toilets can lead to a number of diseases besides an unhealthy environment. In MP, only 37.9 per cent of the respondents clean toilets daily. In Odisha, nearly half of the respondent's clean toilets once in a month and only 9.2 per cent of the people clean toilets daily. In AP, almost half of respondents clean toilets at least once in a week, and 46.4 per cent daily (Figure 6.7).

Figure 6.7

Frequency of Cleaning Toilets

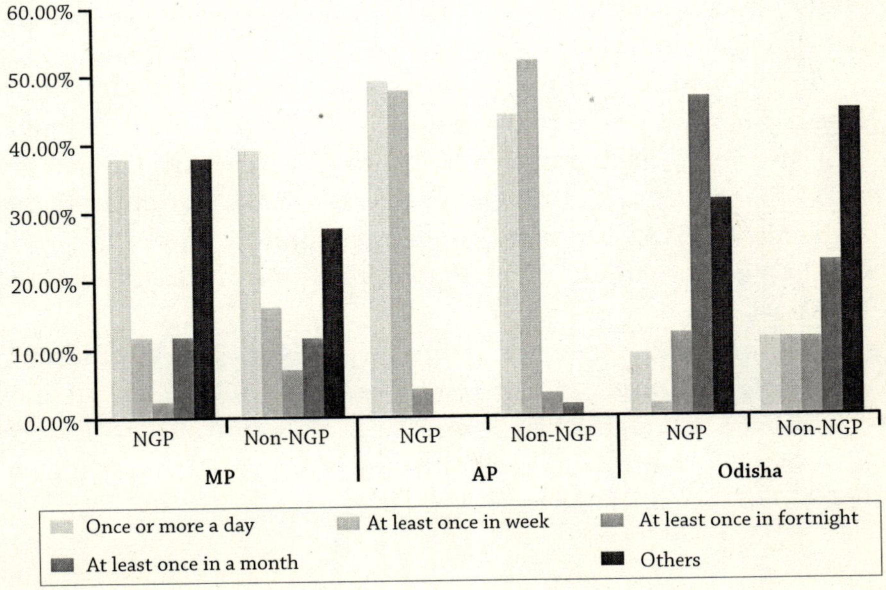

Constraints Related to the Use of In-House Toilet Facility

There are various problems reported regarding the use of toilets within the premises of households. Regarding problems in the use of toilets, more number of respondents have reported problems in MP as compared to Odisha and AP. In MP, the respondents have reported a number of problems like flies and mosquitoes, foul smell, filled up pits etc. Hence, merely having a ready access to toilet facility does not explain the wellness of sanitation facilities. In Odisha, those having in-house toilet facility have reported problems like foul smell, super structure not ensuring privacy, filled up pits (particularly during rainy season in NGP villages), mosquitoes etc., whereas in AP, the problems cited are inadequate water supply, super structure not ensuring privacy, unstable slabs etc.

Practices Related to the Disposal of Children's Excreta

Overall, as far as children's excreta disposal practices are concerned, it can be stated that unhygienic practices are observed more or less in all the three states, with no thought given to the public health consequences. This can be attributed either to the lack of awareness or sheer indifferent attitude on the part of households which needs to be confirmed with a further investigation. A further data break up on this aspect is given as follows.

In MP, regarding the disposal of children's excreta after open defecation, most of the respondents (61.9%) in both NGP and non-NGP areas have mentioned that they throw it on the street. Similarly, in NGP areas, 59.2 per cent of the respondents and most of the respondents (64.7%) in non-NGP villages throw it on the street. In Odisha state, two thirds of the respondents throw children's waste on the street.

While in AP, a majority of the respondents throw children's excreta on the street (73.8%), followed by 'throw in the latrine' (19.3%) and 'left there where it is' (7%). In NGP areas, the percentage of respondents throwing children's waste in the latrine is higher (24.0%) as compared to non-NGP (14.6%) areas.

Personal Hygiene Practices

With regard to personal hygiene practices like hand wash facility near toilets; the situation is better in AP and MP when compared to Odisha. A majority of respondents have hand wash facility near toilets at home in MP. Further, some of them use soap/ash for washing their hands after defecation. In Odisha state, three fourths of the respondents do not have any facility to wash their hands near toilets and as such do not wash hands. This situation is prevalent more in non-NGP villages than in NGP villages. In AP, at the overall level, 98.6 per cent of the respondents have come out with positive responses. However, most of them belong to NGP villages. The respondents having hand wash facilities near toilets are a little higher in NGP villages (98.7%) than those in non-NGP villages (98.4%).

Functioning of the Local Level Institutions

Efficient functioning of the intermediary institutions like VWSC and VHSC is critical to a proper availability and utilisation of water and sanitation facilities among communities and there by improved health conditions of the household members. In AP, although not many respondents are aware of these committees, in some villages, these committees are working through their periodic meetings. In Odisha, one third of the respondents are not aware of the committees and their functioning in both NGP and non-NGP areas. In MP, most of them are not aware of the existence of these committees and their functioning.

In MP, VWSCs do not hold meetings regularly, especially in non-NGP villages, as compared to NGP villages. Many times, minutes of the meetings are not recorded and the issues related to water supply are discussed orally (as revealed during interviews), especially in non-NGP villages. Similarly, many of the villagers are not aware of the existence of VHSC and its functioning at the village level.

In Odisha state, two thirds of the villages have these committees formed and in one third of the villages, respondents are not aware of such committees. Although VWSCs have been constituted in some villages, their functioning is very discouraging as the members are not aware of their duties and powers. This is true in respect of both the NGP and non-NGP villages.

In AP, in both the areas (NGP and non-NGP), only 9 per cent of the respondents are aware of VWSC. In NGP about 15 per cent, and in non NGP villages, 3.3 per cent of the respondents are aware of this committee. As regards the frequency of VHSC meetings, mostly monthly meetings are held in both NGP (66.7%) and non-NGP (83.3%) villages. Maintenance of records of VHSC meetings is done in all the villages of NGP and non-NGP. In NGP villages, the records are maintained mostly by the village sarpanch and other committee members (both together in four villages). But in non-NGP villages, other committee members maintain records in all the villages. VHSCs brings the health issues mentioned in the report to the notice of the relevant health functionaries in five villages in NGP (83.3%) and all the villages in non-NGP (100%).

Conditions of water and sanitation in schools: a comparative status across the three states. In AP, the source of water in most of the schools in non-NGP villages is tap and tank, whereas in NGP villages, schools depend on different sources like tap, tank, public tap, can water and other sources. In MP, the source of water in schools across both the NGP and non-NGP villages is hand pump. In Odisha, borewell is the important source of water in all the schools across both the NGP and non-NGP areas.

In AP, most of the schools have septic tank type latrines in both the NGP and non-NGP areas. Similarly, all the schools in non-NGP villages in MP and also most of the schools in NGP villages have pit latrine. Besides, here piped and septic tank type latrines are also reported. In Odisha, most of the schools in both NGP and non-NGP villages have septic tank type latrines, whereas, a few schools have pit latrines and piped sewerage system.

Morbidity among Children

With the help of a child schedule, the illness data (during the last 12 months) regarding children was collected from the sample households. In AP, more than 77 of the children are reported to have suffered from one or other ailment, while in non-NGP villages, slightly more number of children, 79 are reported to have suffered from illnesses. In MP also, more than half of the children from the surveyed households are reported to have been afflicted with ailments, while slightly more number of children (53.5%) in NGP villages are reported to have been afflicted with ailments. In Odisha, a comparatively less number of children (more than one-fourth) are reported to have suffered from health ailments. As observed in respect of MP and in Odisha also, NGP villages have reported with more number of children with health ailments (37%).

In MP, a relatively more number (139) of children suffering from ailments in NGP areas have been reported as compared to non-NGP areas (131), although the difference is only marginal. In both the areas, a majority of the suffering children are reported to have been taken to private medical practitioners/private hospitals for treatment. In non-NGP areas, 25 diarrheal cases as against 29 in NGP areas have been reported. In both the areas, all the child patients are reported to have got treatment from private hospitals. In non-NGP areas, 5 malaria child patients, and in NGP areas only 2 malaria patients have been reported. In non-NGP areas, 33 children are reported to have suffered from cough, cold and fever as against 30 children in NGP areas. In NGP areas, only 2 cases of dysentery, while only one case of dysentery in non-NGP areas have been reported.

In Odisha, in NGP villages, 74 children are reported to have fallen ill from various water related problems. Among 74 children, 34 are reported to have suffered from various illnesses under the category of waterborne diseases including 13 diarrhoea cases. In non-NGP villages not a single case of diarrhoea, while 7 cases of dysentery in NGP villages and only 4 cases of dysentery in non-NGP villages, have been reported. Only 1 case of malaria in NGP villages and in non-NGP villages not a single case have been reported. In non-NGP areas, 9 children and 13 children in NGP areas are reported to have suffered from cough, cold and fever (multiple health problems).

In AP, in NGP areas, out of 164 children, 75 are reported to have been affected by waterborne diseases, 77 by other diseases which may/may not be related to water and another 12 by other diseases not related to water at all. Out of a total of 75 cases reported suffering from waterborne diseases, 61 come under dysentery, of which 38 are reported to have taken treatment from private clinics. Innon-NGP areas, 95 children out of a total of 187 are reported to have suffered from waterborne diseases, 77 children from dysentery and a majority (58) of them are reported to have approached private clinics for treatment. Three cases of diarrhoea in non-NGP areas and 2 cases in NGP areas of which all the patients are reported to have taken treatment from private clinics. In non-NGP areas, six children and 9 children in NGP areas are reported to have suffered from malaria. In non-NGP areas and NGP areas, 46 and 53 children respectively are reported to have suffered from cough, cold and fever (multiple health problems). A majority of them went to private clinics for treatment (Figure 6.8).

Figure 6.8

Health Problems among Children

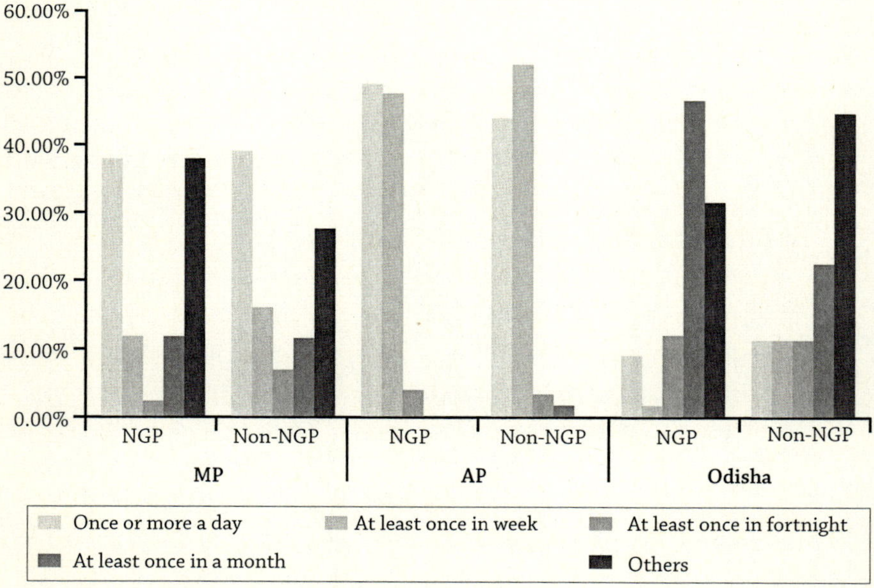

Morbidity among Adults

With the help of a household schedule, the illness data (during last 12 months) regarding adults was collected from the sample households. In AP, around half of the surveyed adults have reported health problems. In non-NGP villages, a slightly more number of adults (51.4%) have reported these problems. In MP, only one-fourth of the surveyed adults have reported health ailments. Proportions of persons with ailments are more or less the same in both NGP and non-NGP areas. In Odisha, comparatively, a less number of adults (around one fifth) have reported health problems and comparatively NGP villages have reported more number of disease affected persons (22.5%).

Regarding health problems among adults in MP, 6 cases in NGP areas and 6 cases diarrhoea in non-NGP areas have been reported. In both the areas, most of the patients are reported to have visited private clinics for treatment. In non-NGP areas, 50 adults are reported to have suffered from cough, cold and fever (multiple health problems) as compared to 51 adults in NGP areas, with a majority of them taking treatment from private clinics. No dysentery cases have been reported in both NGP and non-NGP areas.

In Odisha, 7 cases of dysentery in NGP and 3 cases of dysentery in non-NGP have been reported. In NGP areas, 15 cases of diarrhoea have been reported, while in non-NGP none. In non-NGP areas, 37 adults are reported to have to have suffered from cough, cold and fever (multiple health problems) as against 20 adults in NGP areas, with a majority of them receiving treatment from CHCs (community health centres) and private clinics.

In AP, 47 cases of dysentery in NGP areas and 56 cases of dysentery in non-NGP areas have been reported with most of them in both the areas receiving treatment from private clinics. In NGP, 5 cases of diarrhoea, while only 3 cases in non-NGP areas have been reported. 16 cases of malaria in NGP areas, while 19 cases of malaria in non-NGP areas have been reported. In non-NGP areas, 2 adults are reported to have suffered from cough, cold and fever (multiple health problems) as compared to 2 adults in NGP areas with a majority of them receiving treatment from private clinics (Figure 6.9).

Figure 6.9

Morbidity among Adults

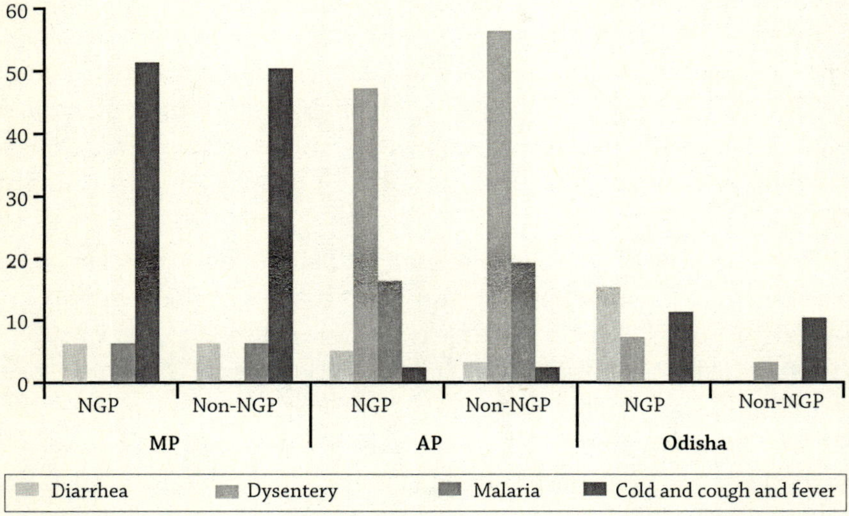

Summary and the Way Forward

The overall situation of water and sanitation status across the three states shows that AP (both NGP and non-NGP) and Odisha (NGP villages only) have a better access to safe drinking water as compared to villages in MP. Even with regard to the quantity of water available, both AP and Odisha are for better when compared to MP, whereas when it comes to the quality (colourless and no hard water) of water, AP state's position is better (pure water for drinking) as compared to Odisha and MP. As regards the availability of toilet facility, AP and MP have performed well when compared to Odisha. Regarding VHSC committees, in terms of conducting of meetings and maintenance of records, the performance of AP state is slightly better than MP and Odisha. Regarding housing, the position of a majority of the households in AP (both NGP and non-NGP villages) is better (people living in *pucca* houses) compared to their counterparts in the other two states (MP and Odisha). Regarding treating of water before drinking, in AP and MP states, the situation is better (both NGP and non-NGP villages) as compared to Odisha.

The overall situation in the study villages across all the three states shows that water and sanitation programmes do not seem to have shown much of influence on the health status of communities. This can be partly explained in terms of the very selection of villages by the government for the NGP award. This award is supposedly awarded to those villages which have achieved a total open defecation-free status including schools and Integrated Child Developments Services (ICDS) centres in the respective villages. Hence, in an ideal situation, NGP villages are expected to be ODF (open defecation free) while maintaining very good sanitation and hygiene conditions along with an assured supply of protected water. But on the ground, as our study reveals, except in the case of a few villages, in all the three states, the selection of villages for NGP has not been based on objective criteria and the whole process may have been driven by extraneous factors (such as political/patronage etc.). As a result of this undesirable process of selection and award of *Puraskar* to villages without properly verifying the achievement of sanitation levels, problems continue to persist in these study villages. And the association or influence of these water and sanitation facilities on the health conditions of households is not strikingly demonstrable in NGP and non-NGP villages across all the three states.

Another important finding of the study is the very ineffective functioning of institutions like VWSCs and VHSCs in most of the villages in both NGP and non-NGP areas across all the three states. In spite of their presence in most of the villages, their actual functioning is not satisfactory. This is amply evident from the number of meetings held as against the specified norms and even if held, the effectiveness of discussions held in the meetings seems to be poor as demonstrated by the records. Our discussions/interviews held with the household respondents during the field work also reveal that in all the villages, many of them are mostly ignorant of the existence/working of these village level institutions. The emphasis given to the working of these institutions in the National Rural Health Mission (NRHM) guidelines has not been effectively translated into practice in all the three study states.

Although NRHM guidelines stress the importance of involving PRIs (panchayat raj institutions) in the management of water and sanitation, in actuality, the involvement is abysmal. Only a few informed *sarpanches/pradhans* in a few villages are actively involved. This despite the 73rd and 74th Constitutional Amendment Acts which make it clear that the provision of water & sanitation facilities falls under the jurisdiction of PRIs and Urban Local Bodies. Hence, there is a need for involving PRIs actively in putting this into practice.

It has been found by the present study that ODF is still being practised which does not augur well for the well-being of children and adults. This could be due to a number of reasons viz.-very poor quality of toilets constructed and hence, using them is not reportedly feasible. An irregular supply of water is a major issue. In addition to these supply side factors, certain cultural beliefs and practices do come in the way of having toilets within the household premises (demand side factors). On the whole, the intensive awareness campaign both by Civil Society Organisation and PRIs—must be accorded a high priority. Otherwise, the poor health outcomes can certainly mar children's future and states need to spend sufficient resources on public healthcare.

An institutional vacuum in terms of a poor functioning of VWSCs and VHSCs can have its toll on the availability and maintenance of the quality of water and sanitation facilities besides ultimately resulting in poor health conditions as observed in the field. A mere formation of these administrative institutions is not a sufficient condition in itself, but making these local structures/institutions accountable to the local people/community.

The study notes that in respect of all the three states, there is a long way to go before attaining a 100 per cent ODF status and better health and hygiene conditions. Policy pointers and lessons learnt:

- Need for strengthening the community toilets is felt in all the states, as most of the individual toilets found are not in a usable condition.
- PRIs and VHSCs need to engage the sanitation staff in maintaining community toilets, on a regular basis (as is the case with the water staff maintenance by PRIs).

- Sustainability of water and sanitation infrastructure is the key and this demands the allocation of committed financial resources.
- A transparent selection process of the NGP villages strictly adhering to Total Sanitation Programme (TSP) norms is needed (a few model villages viz., Hajippally-Mahabunagar, Gangadevipally-Warangal district in Telengana, need to be emulated and scaled up)
- In NGP areas of Odisha, the water table is low; normal toilets have over flowing problem during rainy season. Hence, the government may promote appropriate toilets.
- In all the surveyed states, the functioning of committees (VWSCs and VHSCs) is not satisfactory even wherever committees are found. These are not adequately funded and whatever funds are available are mostly used for spraying bleaching powder and nothing else.
- Except in two NGP villages of Odisha, NGOs (non-governmental organisations) are not found involved in water and sanitation activities. NGOs may be encouraged to promote awareness among people regarding sanitation and hygienic practices along with line departments and PRIs.
- Developing adequate capacities of the local institutions (as it happens with SHGs) in handling water and sanitation issues on a continuous basis is a clear policy imperative.

References

Accountability Initiative (2011). "Total Sanitation Campaign", *Research and Innovation for Governance Accountability, Budget Briefs, Government of India*, 2011-12. New Delhi: Centre for Policy Research.

Alivelu, G., V. Ratna Reddy, P. Bhusen and V. Anitha (2012). "Costs and Service Levels of Water and Sanitation: A Study of Peri-Urban Locations in Andhra Pradesh", *WASHCost-CESS Working Paper* No. 20. Hyderabad.

Annette Prüss-Üstün, Robert Bos, Fiona Gore and Jamie Bartram (2008). *Safer Water, Better Health: Costs, Benefits and Sustainability of Interventions to Protect and Promote Health*. Geneva: World Health Organization.

Barnard, Sharmani, Parimita Routray, Fiona Majorin, Rachel Peletz, Sophie Boisson, Antara Sinha and Thomas Clasen (2013). "Impact of Indian Total Sanitation Campaign on Latrine Coverage and Use: A Cross-Sectional Study in Orissa Three Years Following Programme Implementation", Edited by Abdisalan Mohamed Noor. *PloS One* 8(8): (January). e71438. doi:10.1371/journal.pone.0071438. *http://dx.plos.org/10.1371/journal.pone.0071438*

Brown, Joe, Sandy Cairncross and Jeroen H.J. Ensink (2013). "Water, Sanitation, Hygiene and Enteric Infections in Children", *Archives of Disease in Childhood* (June). doi:10.1136/archdischild-2011-301528. *http://adc.bmj.com/content/early/2013/ 06/11/archdischild-c2011-301528.short*

Bussa, M.T., Mustefa Abdi and Mohammed Ibrahim (2012). *Filtu Water and Sanitation Project: Final Evaluation Report, Ethiopia. http:// www.norad.no/en/tools-and-publications/publications/reviews-from-organisations/publication?key=398482*

Buttenheim, Alison M. (2008). "The Sanitation Environment in Urban Slums: Implications for Child Health", *Population and Environment* 30(1-2): 26-47. doi:10.1007/ s11111-008-0074-9. *http://dx.doi.org/10.1007/s11111-008-0074-9*

Census of India (2011). *Tables on Houses, Households Amenities and Assets Version 2.0.*

Chandrasekhar, V. (2009). "Social Determinants of Health and Health Equity", *Indian Journal of Public Health* 52(2): 79-82. *http://www.ijph.in/show BackIssue.asp?issn=0019-557X;year=2009;volume=53;issue=2;month=April- June#*

Clasen, Thomas, Sophie Boisson, Parimita Routray, Oliver Cumming, Marion Jenkins, Jeroen H.J. Ensink, Matthew C. Feeman, Soosai Peppin and Wolf-Peter Schmidt (2012). "The Effect of Improved Rural Sanitation on Diarrhoea and Helminth Infection: Design of a Cluster-Randomized Trial in Orissa, India", *Emerging Themes in Epidemiology* 9: 7.

Deb, Soumya, Sinjita Dutta, Aparajita Dasgupta and Raghunath Misra (2010). "Relationship of Personal Hygiene with Nutrition and Morbidity Profile: A Study among Primary School Children in South Kolkata", *Indian Journal of Community Medicine* 35(2): 280-84.

Department of Drinking Water and Sanitation. (2010). *Strategic Plan, 2011-2022: Department of Drinking Water and Sanitation—Rural Drinking Water*. New Delhi. *http://ncst.nic.in/writereaddata/linkimages/Strategic Plan_2011_22_Water4619582668.pdf.*

Dobe, Madhumita, A.K. Sur and B.B. Biswas (2011). "Sanitation: The Hygienic Means of Promoting Health", *Indian Journal of Public Health* 55(1): 49-51 (January). doi:10.4103/0019-557X.82557. *http://www.ijph.in/article.asp?issn=0019-557 X;year= 2011;volume=55;issue=1;spage=49;epage =51;aulast= Dobe*

Ganesh, S.R., S. Ganesh Kumar, Sonali Sarkar, Sitanshu Sekhar Kar, Gautam Roy and K.C. Premarajan (2013). "Assessment of Village Water and Sanitation Committee in a District of Tamil Nadu, India", *Indian Journal of Public Health* 57(1): 43-46. doi:10.4103/ 0019-557X.111376. http:// www.ncbi.nlm.nih.gov/pubmed/23649144.

George, Benny (2009). "Nirmal Gram Puraskar: A Unique Experiment in Incentivising Sanitation Coverage in Rural India", *International Journal of Rural Studies* 16(1): 1-6. *http://www.vri-online.org.uk/ijrs/April2009/incentivising*

Gireesan, K. and Jos Chatukulam (2007). "Total Sanitation Campaign and Nirmal Gram Puraskar: An Overview of Four Indian States", *Gandhi Marg* 29(2):157-78.

Gleick, Peter H. (2002). *Dirty Water: Estimated Deaths from Water-Related Diseases 2000-2020*. Oakland, California. *http://www.pacinst.org/wp-content/uploads/ sites/21/2013/02/water_related_deaths_report3.pdf*

Government of India (GoI) (2012). *Nirmal Gram Puraskar Guidelines*. New Delhi: Ministry of Drinking Water and Sanitation.

Harris, Gardner (2014). "Children's Malnutrition Linked to Poor Sanitation: India's Stunting Problem Represents the Largest Loss of Human Potential in Any Country", Article in *Deccan Herald-Panorama-INYT* (*International New York Times*), 15 July.

Henry, F.J., S.R.A Huttly, Y. Patwary and K.M.A Aziz (1990). "Environmental Sanitation, Food and Water Contamination and Diarrhoea in Rural Bangladesh", *Epidemiology & Infection* 104(02): 253-59.

Joshi, Ashish and Chioma Amadi (2013). "Impact of Water, Sanitation and Hygiene Interventions on Improving Health Outcomes among School Children", *Journal of Environmental and Public Health* (January): 984626. doi:10.1155/2013/ 984626. *http://www.readcube.com/articles/10.1155/2013/984626.*

Kondepati, Rajendra (2013). *Demand Creation for Rural Sanitation in Andhra Pradesh, India*. March. *http://papers.ssrn.com/abstract=2271966*

Kurian Baby, V. and V. Ratna Reddy (2012). "WASH Security in India: Can the New Policy Guidelines Deliver Critical Assessment and Operationalization of 2010 Guidelines", *WASHCost CESS Working Paper* No.21. Hyderabad.

Lin A, B.F. Arnold, S. Afreen, R. Goto, T.M.Huda, R. Haque, R. Raqib, L. Unicomb, T. Ahmed, J.M.Colford and S.P.Luby (2013). "Household Environmental Conditions are Associated with Enteropathy and Impaired Growth in Rural Bangladesh", *The American Journal of Tropical Medicine and Hygiene* 89(1): 130-37.

Matewos Tera Bussa, Mustefa Abdi, Mohammed Ibrahim (2011). *Filtu Water and Sanitation Project: Final Evaluation Report, Ethiopia. http:// www.norad.no/en/tools-and-publications/publications/reviews-from-organisations/publication?key=398482*

Minh, Hoang Van and Hung Nguyen-Viet (2011). "Economic Aspects of Sanitation in Developing Countries", *Environmental Health Insights* 5: 63-70. doi:10.4137/ EHI.S8199. *www.la-press.com/economic-aspects-of-sanitation-in-developing- countries-article-a2871*

Ministry of Drinking Water and Sanitation (2013). *Outcome Budget*. New Delhi. Retrieved: *http://mdws.gov.in/hindi/sites/upload_files/ddwshindi/files/pdf/ Outcome_Budget_2013_14.pdf*

Ministry of Health and Family Welfare (n.d.) *National Health Mission (NHM), Goals*. Government of

India.

———. (n.d.). *National Rural Health Mission: Meeting People's Health Needs in Rural Areas—Framework for Implementation 2005-2012*. Government of India.

Orlove, Ben and Steven C. Caton (2010). "Water Sustainability: Anthropological Approaches and Prospects", *Annual Review of Anthropology* 39(1): 401-15 (September). doi:10.1146/annurev.anthro.012809.105045. *http://dx.doi.org/ 10.1146/annurev.anthro.012809.105045*

Pardeshi, Geeta, Avinash Shirke and Minal Jagtap (2008). "SWOT Analysis of Total Sanitation Campaign in Yavatmal District of Maharashtra", *Indian Journal of Community Medicine* 33(4): 255-59. *http://www.ncbi.nlm.nih.gov/pmc/ articles/PMC2763692/*

Planning Commission (2010). *Evaluation Study on Rajiv Gandhi National Drinking Water Mission (RGNDWM)*. New Delhi. Retrieved (*http://planningcommission.nic.in/ reports/peoreport/peoevalu/peo_rgndwm.pdf)*.

Prüss-ustun, A., David Kay, Lorna Fewtrell and Jamie Bartram (2002). "Environmental Health Perspectives", *Environmental Health Perspectives* 110(5): 537-42. *http://www.who.int/quantifying_ehimpacts/global/en/ArticleEHP05(2002).pdf*

Rama, Chandrudu, M.V, R. Subramanyam Naidu, Safa Fanaian and Radha Shree (2012). "Qualitative Challenges in Improving Performance of Water Points: Insights from Community Based Performance Assessment of Water Points in Andhra Pradesh", *WASHCost CESS Working Paper* No.19. Hyderabad.

Rammohan Rao, M., M. Venkataswamy, C. Ramachandraiah and G. Alivelu (2012). "Urban Water Supply and Sanitation in Andhra Pradesh", *WASHCost CESS Working Paper* No. 16. Hyderabad.

Ratna Reddy.V., M.S. Rama Mohan Rao and M. Venkataswamy (2010). "Slippage: The Ban of Rural Drinking Water Sector (A Study to Extent and Causes in AP)", *WASH Cost CESS working paper* no-6. Hyderabad.

Ray, S.K. (2009). "Importance of Hand Washing Practices for Prevention of Communicable Diseases: A Forgotten Issue", *Indian Journal of Public Health* 53(2): 63-66.

Ray, Sandip Kumar, Forhad Akhtar Zaman and Nasrin Banu Laskar (2010). "Hand Washing Practices in Two Communities of Two States of Eastern India: An Intervention Study", *Indian Journal of Public Health* 54(3): 126-30. *http:// www.ijph.in/article.asp?issn=0019-557X;year= 2010; volume=54;issue=3; spage=126;epage=130;aulast=Ray*

Sharma, S.K. (2007). "Recent Advances Made in Achieving Basic Sanitation and Hygiene in India", *GEWASSERSCHUTZ WASSER ABWASSER* 206: 11.

Snehalatha, M., M. Venkataswamy, D. Sirisha, V. Anitha and P. Busenna (2012). "Looking Beyond Capital Costs: Life Cycle Costing for Sustainable Service Delivery—A Study from Andhra Pradesh, India", *WASHCost CESS Working Paper* No.17. Hyderabad.

Spears, Dean, Arabinda Ghosh and Oliver Cumming (2013). "Open Defecation and Childhood Stunting in India: An Ecological Analysis of New Data from 112 Districts (Edited by Vishnu Chaturvedi)", *PloS One* 8(9): e73784 (January). doi:10.1371/journal.pone.0073784. *http://dx.plos.org/10.1371/journal.pone. 0073784*

Supply, Department of Drinking Water and Development. (2010). *National Rural Drinking Water Programme. http://www.gwssb.org/pdf/program/NRDWMP/RuralDrinkingWater_2nd April.pdf*

Suresh Reddy, B. and M. Snehalatha (2011). "Sanitation and Personal Hygiene: What does it Mean to Poor and Vulnerable Women", *Indian Journal of Gender Studies* 18(3)- 381-404(c). sage Publications.

Wagner, John R. (2012). "Water and the Commons Imaginary", *Current Anthropology* 53(5): 617-41 (October). doi:10.1086/667622. *http://www.jstor.org/stable/ 10.1086/667622*

Water Aid (2007). *Community Led Total Sanitation (CLTS) An Evaluation of the WaterAid's CLTS Programme in Nigeria.*

Watkins, Kevin, Liliana Carvajal, Daniel Coppard, Ricardo Fuentes, Arunabha Ghosh, Chiara Giamberardini, Claes Johansson Papa Seck, Cecilia Ugaz and Shahin Yqub (2006). "Beyond Scarcity: Power, Poverty, and the Global Water Crisis", *Human Development Report* 2006. The United Nations Development Programme(UNDP).

Young, J., and Briscoe, B. (1988). "A Case-Control Study of the Effect of Environmental Sanitation on Diarrhoea Morbidity in Malawi", *J Epidemiol Community Health* 42(1): 83-88. *http://www.ncbi.nlm.nih.gov/pmc/articles/PMC1052686/*